Risk in the Technological Society

AAAS Selected Symposia Series

Risk in the Technological Society

Edited by Christoph Hohenemser and Jeanne X. Kasperson

Routledge
Taylor & Francis Group

LONDON AND NEW YORK

AAAS Selected Symposium **65**

First published 1982 by Westview Press

Published 2018 by Routledge
52 Vanderbilt Avenue, New York, NY 10017
2 Park Square, Milton Park, Abingdon, Oxon OX14 4RN

Routledge is an imprint of the Taylor & Francis Group, an informa business

Copyright © 1982 by the American Association for the Advancement of Science

Library of Congress Cataloging in Publication Data
Main entry under title:
Risk in the technological society.
     (AAAS selected symposium ; 65)
     "Based on a symposium which was held at the 1980 AAAS National Annual Meeting in San Francisco, California, January 3-8."
     Includes index.
     1. Technology assessment--Congresses. 2. Risk--Congresses. I. Hohenem-
ser, Christoph. II. Kasperson, Jeanne X. III. American Association for the Advancement of Science. IV. Series.
T174.5.R57   303.4'83   81-14745
ISBN 0-86531-316-4   AACR2

ISBN 13: 978-0-367-28612-5 (hbk)

ISBN 13: 978-0-367-30158-3 (pbk)

About the Book

In this book, representatives of government, industry, universities, and public interest groups consider the emerging art of risk assessment and discuss the issues and problems involved. They look at two failures in technological risk management--Three Mile Island and Love Canal; examine the dimensions of technological risk; tackle the difficult question of how safe is "safe enough"; and offer a set of research priorities.

About the Series

The *AAAS Selected Symposia Series* was begun in 1977 to provide a means for more permanently recording and more widely disseminating some of the valuable material which is discussed at the AAAS Annual National Meetings. The volumes in this *Series* are based on symposia held at the Meetings which address topics of current and continuing significance, both within and among the sciences, and in the areas in which science and technology impact on public policy. The *Series* format is designed to provide for rapid dissemination of information, so the papers are not typeset but are reproduced directly from the camera-copy submitted by the authors. The papers are organized and edited by the symposium arrangers who then become the editors of the various volumes. Most papers published in this *Series* are original contributions which have not been previously published, although in some cases additional papers from other sources have been added by an editor to provide a more comprehensive view of a particular topic. Symposia may be reports of new research or reviews of established work, particularly work of an interdisciplinary nature, since the AAAS Annual Meetings typically embrace the full range of the sciences and their societal implications.

WILLIAM D. CAREY
Executive Officer
American Association for
the Advancement of Science

Contents

PART 4. AGENDA FOR RESEARCH

About the Editors and Authors

Christoph Hohenemser *is professor of physics and co-founder and chairman of the Program on Science, Technology and Society at Clark University. An experimental physicist working in magnetism and materials science, he has been active in hazard management and risk analysis for seven years, and he has written on nuclear power policy, arms control, and transportation risks. He is coeditor of* Technological Hazard Management *(with R.W. Kates; Oelgeschlager, Gunn & Hain, forthcoming).*

Jeanne X. Kasperson *is a research librarian at the Center for Technology, Environment and Development at Clark University. She is the author of articles on the accident at Three Mile Island and is coeditor of* Water Re-Use and the Cities *(with R.E. Kasperson; University Press of New England, 1977).*

Joseph F. Coates *is president of J.F. Coates, Inc., a policy analysis consulting firm in Washington, D.C. A chemist by training, his major current interest is planning for the future, with emphasis on the impact of technology on society. He is a former assistant to the director of the Office of Technology Assessment of the U.S. Congress and program manager for technology assessment at the National Science Foundation.*

Vincent T. Covello, *program manager in risk analysis at the National Science Foundation, is a specialist in policy analysis, technology assessment, risk analysis, and evaluation research. He has published on the evaluation of federal research and development programs and on the use of research in policy decision-making.*

Paul F. Deisler, Jr., *a chemical engineer by training, has spent most of his career in various research and operating positions within the Shell Oil Company. He is now Shell Oil Company's Vice President for Health, Safety and Environment.*

Lois R. Ember, *a specialist in environmental science and health, is an associate editor for* Chemical & Engineering News. *Her writing has earned her awards from the Society of Technical Communication and the National Association of Recycling Industries.*

Baruch Fischhoff, *a research associate at Decision Research (a branch of Perceptronics, Inc.) in Eugene, Oregon, has specialized in risk analysis and decision-making and judgment under uncertainty. He has written numerous articles on risk perception, surveying of public values on technology management, and on the quality of scientific judgment, and he is the author of* Acceptable Risk *(with S. Lichtenstein, P. Slovic, S. Derby and R. Keeney; Cambridge University Press, 1981).*

Gio Batta Gori, *a microbiologist by training, is vice president of the Franklin Institute and director of the Institute's Policy Analysis Center. A former deputy director of the Division of Cancer Cause and Prevention at the National Cancer Institute, he has initiated studies on occupational and nutritional cancer, smoking and disease, and related regulatory policies. He currently directs a program to provide independent evaluation of health hazards and the costs/benefits of federal regulatory policy.*

Stanley M. Gorinson *was chief counsel of the President's Commission on the Accident at Three Mile Island and is presently Chief, Special Regulated Industries Section, U.S. Department of Justice, Antitrust Division. He has written reports on the Three Mile Island nuclear reactor accident; the U.S. Nuclear Regulatory Commission; the role of the managing utility and its suppliers; and on emergency preparedness.*

Roger E. Kasperson, *professor of geography and government and director of the Center for Technology, Environment and Development, Clark University, is a specialist in risk assessment, workplace hazards, nuclear power policy issues, and radioactive waste management. He edited* Equity Issues and Radioactive Waste Management *(with R. Kates; Oelgeschlager, Gunn & Hain, forthcoming) and* Water Re-Use and the Cities *(with J. Kasperson; University Press of New England, 1977).*

Robert W. Kates *is professor of geography and research professor in the Center for Technology, Environment and Development, Clark University. His research interests are in risk assessment and management of hazards, climate impact assessment, and the theory of the human environment, and he has published numerous books and articles in these fields,*

including Technological Hazard Management *(edited with C.*
Hohenemser; Oelgeschlager, Gunn & Hain, forthcoming) and
Risk Assessment of Environmental Hazard *(Wiley International,*
1978).

William J. Librizzi, Jr., *a civil and sanitary engineer*
by training, is director of the Surveillance and Analysis
Division (Region 6) at the U.S. Environmental Protection
Agency. As chief of the Office of Toxic Substances, he was
responsible for EPA's response to the crisis at Love Canal,
including the initial field and analysis activities and
intergovernmental coordination.

Sarah Lichtenstein *is a research associate at Decision*
Research (a branch of Perceptronics, Inc.) in Eugene, Oregon,
and adjunct professor of psychology at the University of
Oregon. Her research interests are judgment and decision-
making, decision aids, and risk assessment, and she has
published extensively in these fields.

Joshua Menkes, *a group leader in technology assessment*
and risk analysis, Division of Policy Research and Analysis,
National Science Foundation, is a specialist in technology
assessment and systems analysis. He has written extensively
on the interaction of technology and society, and he is a
member of the advisory board of Technological Forecasting and
Social Change.

Murdo Morrison *is a doctoral candidate and research*
assistant in geography at Clark University's Center for
Technology, Environment and Development. He is currently
involved in research on hazard management, examining patterns
of control responses to varying hazards.

David Okrent *is a professor of engineering and applied*
science at the University of California, Los Angeles, and is
a member of the National Academy of Engineering. A special-
ist in fast reactor technology, nuclear reactor safety, and
societal risk, he is the author of Nuclear Reactor Safety
(University of Wisconsin Press, 1981).

Russell W. Peterson *is president of the National Audubon*
Society. A chemist by training, he has served as chairman
of the President's Council on Environmental Quality, director
of the U.S. Congress Office of Technology Assessment, governor
of the state of Delaware, and a member of the President's
Commission on the Accident at Three Mile Island.

Richard C. Schwing *is a senior staff research engineer*
in the Societal Analysis Department of General Motors Re-

search Laboratories. A specialist in thermodynamics, chemical kinetics, and nuclear engineering, he has published on risk analysis, decision theory, benefit-cost analysis of automotive emission reductions, and air pollution epidemiology. He is the coeditor of Societal Risk Assessment: How Safe Is Safe Enough? *(with W.A. Albers, Jr.; Plenum, 1980).*

Paul Slovic *is a research associate at Decision Research (a branch of Perceptronics, Inc.) in Eugene, Oregon. His fields of specialization are judgment, decision-making and risk assessment, and he has coauthored several books and written numerous articles on this topic. He serves on the editorial boards of the* Journal of Experimental Psychology, Organizational Behavior and Human Performance, *and* Risk Analysis.

Chauncey Starr *is vice chairman of the Electric Power Research Institute. A specialist in atomic energy and risk analysis, he has served on numerous committees and panels on these topics, including the President's Task Force on Science Policy and the President's Energy Advisory Committee. Among his many publications are* Science, Technology and the Human Prospect *(edited with P. Ritterbush; Pergamon, 1980),* Current Issues in Energy *(Pergamon, 1979), and "Philosophical Basis for Risk Analysis" (with R. Rudman and C. Whipple) in the* Annual Review of Energy *(Vol. 6, 1976).*

Chris Whipple, *a mechanical engineer, is a technical manager in the Energy Study Center, Electric Power Research Institute. A member of the Utilization Advisory Committee, NSF Project on the Societal Management of Technological Hazards, he has taught courses on risk-benefit analysis and is coeditor of* Energy and the Environment: A Risk-Benefit Analysis *(with H. Ashley et al.; Pergamon, 1977).*

Acknowledgments

We thank each of the contributors for being patient
with our quibbles and exercise of editorial license. We
are indebted to Anne Hohenemser, Roger Kasperson, and
Robert Kates for their critical reading and helpful comments on portions of the manuscript. And we are especially
thankful to Joan McGrath, Deborah Morin, and Nancy O'Connor
for their skill and professionalism in typing camera-ready
copy.

Christoph Hohenemser
Jeanne X. Kasperson
Worcestor, Mass.

Christoph Hohenemser, Jeanne X. Kasperson

1. Introduction

Flood, drought, famine, and infectious disease, all of them natural hazards, were once the principal hazards faced by society. Today they have been replaced by hazards arising from technology.[1] Though the benefits of technology are widely acknowledged, we are bombarded daily by media accounts describing new, previously unsuspected technological threats to human health and well-being. It is believed by some that we have reached a state of crisis, that some technologies threaten survival, and that for many others the benefits no longer outweigh the costs.

Ironically, technology is in many ways its own worst accuser. Some hazards are detected only because of our technical ability to track minute concentrations of toxic substances; and other hazards are recognized because of our technical ability to imagine and describe theoretically a range of technological catastrophes.

Few media accounts fail to put at least some of the blame for technological failures on industry or government. The reader, listener, or viewer thus receives the impression that technology managers have erred, and may have deliberately sought private profit at public expense, or succumbed to powerful special interests. Rarely are the problems described in detail sufficient to indicate that solutions may not be easy, that science provides uncertain answers, or that one person's risk implies another's benefit. In effect "our private capacity to generate hazards to health has outstripped our public ability to evaluate and control hazards."[2]

The conflicts that underlie the public discussion of hazards include conflicts about facts, conflicts about perception, conflicts about risks versus benefits, and conflicts driven by divergent views about individual versus societal responsibilities. In many cases it is unclear to the very proponents of particular views what their implicit assumptions are.

For example, most members of the American public consider it self-evident that smoking and automobile seatbelt use should be matters of individual choice. Many Americans are also appalled by the extraordinary cost of health care, particularly for sufferers of chronic diseases and permanent major disabilities. Yet few people realize that smoking and auto accidents account for 350,000 deaths annually[3,4] and directly consume a major portion of the high cost of health care.

The importance of hidden implicit assumptions comes home in the juxtaposition of two newspaper accounts of two different, yet ironically complementary, citizen protests against the siting of electric power plants.[5] One story details the efforts of a middle-aged farming couple, the Shadises, to close down Maine Yankee, an 840-Megawatt nuclear power plant in Wiscasset, Maine, two miles from their dairy herd. The second story describes a neighborhood protest by Bostonians seeking to stop Harvard University's new oil-fired diesel cogeneration plant because of the air pollution it will produce. In each case, the opposition wants to keep power plants out of its "own back yard"-- all the while showing little concern for the other "back yard" where the power is generated.

Along with a certain myopia in relation to total technological systems, it seems that the public has become more demanding. Thus, public health standards continue to become more stringent. Official assurances of safety or "no immediate danger" are met, especially in the aftermath of Three Mile Island and Love Canal (see Part 2, this volume), with suspicion, skepticism, and outright disbelief. This does not necessarily imply a crisis of public confidence in technology per se, but rather a growing doubt among a significant portion of the general public that technology is going to be managed adequately.[6]

The issues that confound the handling of technological risk have become the concern of a massive regulatory apparatus that operates at all levels of government and reaches deeply into the activities of industry and consumers. One recent study cited 12 major federal regulatory agencies and 179 laws concerned with the management of technological risk.[7] Annual expenditures by government are estimated at more than $30 billion per year, and expenditures by government and the private sector combined may be as high as $130 billion (1979), or about 5% of the gross national product.[8]

Much of the regulatory process involves narrow issues, hammered out through long, tedious processes which are constrained by the uncertainties and fundamental assumptions of the enterprise. Thus the Environmental Protection Agency (EPA) is involved in a continuing battle to improve air quality, by setting attainment standards, regulating sources, and occasionally ordering specific action by local governments. On one side the logic of its action is constrained by science, according to which nonacute health effects of air pollution are ill-defined at best.[9] On the other side, the scope of its action is constrained by politics and law, according to which it can not deal with the ultimate causes of air pollution risks, such as the level of human wants and the choice of technology for achieving them. As a result the EPA and similar agencies seek compromises and narrow technological fixes, which leave both consumers and affected industries dissatisfied, and which in a holistic view of the world may involve a nonoptimal use of resources.

In some cases, government and the public make demands for risk regulation that are close to contradictory. Consider the Food and Drug Administration (FDA), charged with safeguarding food and regulating additives to food. Its actions are subject to the Delaney Amendment, a congressional action that stipulates that additives known to be carcinogens must be excluded from food. "Known carcinogens" are almost wholly defined by high-dose animal experiments, and extrapolation to humans is made on the conservative assumption that high dose animal carcinogenesis implies potential human cancer (see Chapter 11). The logic of absolute prohibition, which sounds quite reasonable at first, begins to unravel when it is realized that some animal carcinogens occur as natural substances in food and food processing. Since the enactment of the Delaney amendment in 1959, the growing technical ability to detect traces of chemicals has made the meaning of a zero threshold increasingly questionable; for if it means anything, "zero threshold" means "not detectable." The logic of absolute prohibition unwinds completely with the realization that some animal carcinogens have beneficial functions for which no real substitutes exist.

In response to this and similar situations, regulators and researchers have begun to address the problem of risk evaluation in ways that are generic rather than case-specific. A common thread of such approaches has been the desire to conduct "comparative risk assessment," a process that implies a broad range of activities, including an evaluation of the scientific basis of risk, the social context of technology choice, the dimensions of benefits arising therefrom, and the social dimensions of risk consequences. Not surprisingly, the risk analysts who are beginning to work in government, industry, and the universities come from many branches of natural and social sciences and frequently bring with them their respective disciplinary traditions. These impinge crucially on the very definition of "risk," which is no mere matter of semantics, since it determines what will be studied, and what will not.

A number of physicists and economists have defined risk simply as the per capita frequency or probability that a particular result (e.g. an untimely death) should occur. They have further proposed that the central question of risk assessment is a compilation that expresses all relevant risks in terms of such numbers.[10,11] Although such compilations allow gross scaling of risky technologies and activities and certainly permit risk comparisons, they fail in a fundamental way because they do not reflect other dimensions of hazardousness that may have equal or greater social value (see Chapter 9). In contrast, a number of social scientists have been at work elucidating the complex, multivariate and subjective character of risk judgments. One of their techniques has been to question ordinary people in order to determine the cognitive content of subjective risk. Using this approach, Slovic and his associates (Chapter 10) have found that people rate risks differently when an event kills many people simultaneously rather than one at a time; when activities and technologies are voluntary rather than involuntary; or when they are new rather than old.

Given the divergent views of the meaning of "risk," it should not be surprising that "risk" assessment is a pursuit for which basic goals and definitions are widely debated and not easily agreed upon. Risk assessment is, in short, a field in its infancy.

Beyond defining and measuring risk, an important question is which risks are acceptable? Or equivalently, how safe is safe enough?[12] Implicit in this question is the assumption that a risk-free environment is an elusive goal. In deciding on which risks are acceptable, agreement on fundamental approaches is even less well established than in the case of risk definition and measurement. As in the case of defining the concept of "risk" in the first place, the underlying difficulty has to do with incorporation of multiple human values. This includes the value to be placed

Acceptable risk has been approached through a variety of principles and practice, including: (1) setting quantitative risk standards above which risks are deemed unacceptable; (2) comparing risks to benefits in commensurate terms, and demanding that benefit/risk ratios exceed unity; (3) comparing cost-effectiveness of various risk control strategies in units of cost per life saved or cost per year of longevity; and (4) defining rules of aversion for those cases where negligible benefits accrue to risk taking. As discussed in detail in Chapters 12-16, each of these approaches has its own uses, advantages, and problems, not the least of which is that they are all based in one way or another on a rather narrow definition of risk as "conditional probability of harm."

Risk measurement and evaluation thus concoct a rich and often inconsistent brew. Far from resolving problems of risk management by society, particularly government, the difficulties we have mentioned translate into a number of generic issues that block progress at present. In recent work, the groups at Clark University and Decision Research have identified seven such issues:[13]

Incomplete knowledge. Characteristically, hazard managers would like to assume that causality is or will soon be defined. Yet current knowledge is often insufficient. We know, for example, that the burning of fossil fuel has its effect on climate. We know there are risks, but we cannot say much more and are very far indeed from formulating policy.

Foregoing benefits. Controlling risks has impacts on the benefits of technology. Benefits are often shared by people other than those exposed to risk. Benefits tend to be as clear and tangible as risks are ambiguous and elusive. All this makes for inevitable conflict, rancorous debate, and in the end, ineffective societal action. Examples of this problem are pollution of urban air by automobiles, acid precipitation attributable to the burning of fossil fuels, and the catastrophic risks of nuclear power. For each, effective risk control involves serious impacts on massive benefits, with the result that satisfactory risk management remains an elusive goal.

A limited capacity to react. A myriad of risks confronts us. The list is much longer than our strand of worry beads or our physical and financial capability to respond. There are, for instance, 32,000 chemical substances already in commerce,[14] and 2,400 of those substances may be causing cancer in the workplace.[15] Despite the recent

legislative effort to tackle this problem (i.e. The Toxic
Substances Control Act), we are far indeed from a solution.

Perception of risk. Risk as perceived by people can
vary widely from risk as defined by statistics or models.
This adds conflict to hazard management. For example,
according to the work of Slovic et al. (Chapter 10), people
judge the risk of auto accidents as equal to that of nuclear
power, even though most experts would rank them orders of
magnitude apart.

Value tradeoff. Hazard management inevitably confronts
difficult value decisions, such as tradeoff between present
and future lives. Here as in the case of perceived risk,
conflict ensues with no widely accepted way of resolving the
issues. An excellent example is the problem of nuclear
waste storage which remains, despite a number of possible
technical solutions, "unsolved" in a value sense.[16]

Institutional weaknesses. Institutions, whether regulatory agencies, congressional committees, or public interest groups, grow helter-skelter. Their scopes and programs
are often narrow, time-bound, and dominated by special interests. This prevents taking a sufficiently broad view, and
may lead to misguided hazard management. A good example is
the recent program by the Department of Transportation to
widen the shoulders of interstate highways at a cost of $6
million per life saved. This program was pushed intensively
despite the Department's own research report[17] which identified many more effective ways of spending a given highway
safety dollar (see Chapter 16).

Creation of new hazards. Risk management often creates
new risks. The result may be worse than would have been the
case without societal intervention. Provision of free or in-
expensive driver education to high school students, for
instance, increases the overall accident rate, because such
programs accelerate entry of inherently high-risk drivers
into the driving population.[18] Equivalently, flood control
programs have historically increased total flood damage
because an increased perception of safety encouraged more
intensive settlement on floodplains.[19]

It is our belief that the generic problems of hazard
management are rooted in the causal structure of hazard
(see Chapter 9). At the same time, the results observed
in particular cases depend strongly on the character of
institutions, laws, and perceptions of society. To unify
the research on risk measurement, risk evaluation, and risk
management, a consistent theory appears highly desirable.

Such a theory should begin with unambiguous definitions,
describe what is possible and what is not, what can be done
and what cannot, all in a manner that is consistent with the
laws of nature, the characteristics of human perception and
behavior, and the constraints of our institutions. As our
brief introduction to the issues has shown, such a theory
does not now exist.

While the reader waits for the theory to develop, this
volume offers a modest contribution to a growing literature
on technological risk.[20] Some of the papers for the volume
were drawn from a symposium, organized by Joseph F. Coates,
held at the 1980 Annual Meeting of the American Association for the Advancement of Science. Other papers are
selected from previously unpublished sources and from recent
journal articles. While no claim is made that the collection is any way comprehensive or representative of the field,
it is hoped that the reader will find here some provocative
new contributions by 21 authors, many of whom have made
significant past contributions to the field.

Part 1 is intended to motivate the reader's concern
through concrete examples that illustrate a number of the
generic problems of risk management that we have discussed.
Each of the five contributions, but particularly the first
by Coates, suggests why our present institutional structures
are inadequate to the task at hand. The papers by Peterson
and Gorinson focus on specific institutional lessons to be
learned from the nuclear reactor accident at Three Mile
Island; and the contributions by Librizzi and Ember describe
institutional and individual responses to the still evolving
waste-dumping episode at Love Canal.

Part 2 provides a brief description of the fundamental
structure of technological risk. Hohenemser, Kasperson, and
Kates estimate the size of the hazard burden, discuss causal
structure, and offer a causal taxonomy of hazards. Slovic,
Fischhoff, and Lichtenstein review the surprising and sometimes puzzling characteristics of risk perception. Gori,
while addressing the regulation of carcinogenic hazards,
provides a detailed review of the adequacy of available data
and proposes an alternative taxonomy based on benefit classification.

Part 3 deals with the question of acceptable risk and
reflects fairly accurately the lack of coherence that
currently afflicts that question. Three papers by Starr
and Whipple, Okrent, and Deisler address the issue of quantitative risk standards and strongly advocate their use. A
fourth, by Schwing, describes how cost-effectiveness of risk

control is best evaluated in terms of increased longevity,
not lives saved, all on the assumption that quantitative
data of a suitable kind are available.

Part 4 concludes the volume with two contributions on
future research agendas. The first, by Menkes and Covello,
reflects current thinking in the National Science Foundation's Program on Technology and Risk Assessment (TARA).
The second, by Kasperson and Morrison, offers an international perspective, originally based on a series of proposals prepared for the Swedish government.

Except for some of the case study material in Part 1,
and some commentary by Gori (Chapter 11), our contributors
do not directly approach the problem of what the risk manager and regulator must do, or for that matter, how risk
regulation ought to be organized. Yet there is plenty here
that makes clear why life as a risk manager will be confounded by contradictory demands, inconclusive evidence,
ambiguous criteria, and just general misery. Consequently,
one may confidently expect that this volume, as all messengers
of bad news, will be received with little pleasure by the
soldiers at the front. Nevertheless, it is hoped that
there is sufficient insight in some of the material presented here to cause some actors in the real-life drama of
risk management to think more clearly and act with a better
understanding of the job that confronts them. It is also
hoped that the general reader will come away with a better
understanding of why society can be expected to struggle
for some time with the evaluation and management of technological risk; and that particularly for a democratic society,
it is essential to find better ways of incorporating the
views of ordinary citizens into the decision-making process.

References and Notes

1. R.C. Harriss, C. Hohenemser, and R.W. Kates, "Our
    Hazardous Environment," *Environment,* 20, no. 7 (Septem-
    ber, 1978), 6-13, 38-41.

2. E.M. Kennedy, "Risk/Benefit Decisions in the Regulatory
    Environment," in *Risk/Benefit Decisions and the Public
    Health* (Proceedings of the Third FDA Science Symposium,
    15-17 February 1978; FDA 80-1069), ed. J.A. Staffa
    (Office of Health Affairs, Food and Drug Administra-
    tion, Rockville, MD., 1980), p. 16.

3. *Accident Facts 1979* (National Safety Council, Chicago,
    1979).

4. U.S., Office of Smoking and Health, *Smoking and Health: A Report of the Surgeon General* (DHEW pubn. no. (PHS) 79-50066; Dept. of Health Education, and Welfare, Rockville, MD., 1979).

5. D. Goodman, "Their activism still blooms"; J. Ackerman, "Vision for energy stalled in reality," *Boston Globe,* Metro/Region section, 27 April 1980, p. 41.

6. N. Ashford, Discussion in U.S. Congress, House, Committee on Science and Technology, Subcommittee on Science, Research and Technology, *Risk/Benefit Analysis in the Legislative Process* (Joint hearings before the Subcommittee...and the Subcommittee on Science, Technology and Space of the Committee on Commerce Science and Transportation, U.S. Senate, and Congress/Science Forum with the AAAS; 96th Cong., 1st sess., 24, 25 June 1979; no. 71), p. 54.

7. B.B. Johnson, "Congress as Technological Hazard Manager: Analysis of Legislation on Technological Hazards, 1957-1978," Ph.D. dissertation, Graduate School of Geography, Clark University, Worcester, Massachusetts, 1980.

8. J. Tuller, "Technological Hazards: The Economic Burden," in *Technological Hazard Management,* ed. R.W. Kates and C. Hohenemser (Oelgeschlager, Gunn, and Hain, Cambridge, MA., forthcoming).

9. L.B. Lave and E.P. Seskin, *Air Pollution and Human Health* (Johns Hopkins for Resources for the Future, Baltimore, 1977).

10. B.L. Cohen and I. Lee, "A Catalog of Risks," *Health Physics,* 36 (June, 1979), 707-722.

11. R. Wilson, "The Risks of Daily Life," *Technology Review,* 81 (February 1979), 40-46.

12. *Societal Risk Assessment: How Safe is Safe Enough?,* ed. R.C. Schwing and W.A. Albers (Plenum Press, New York, 1980).

13. B. Fischhoff, C. Hohenemser, R.E. Kasperson, and R.W. Kates, "Handling Hazards," *Environment* 20 no. 7 (September, 1978), 16-20, 32-37.

14. U.S., Environmental Protection Agency, Office of Toxic Substances, *Candidate List of Chemical Substances,*

vols. 1-3 (Washington, 1977); *Addenda 1 and 2* (Washington, 1978).

15. National Institute for Occupational Safety and Health, *Suspected Carcinogens: A Subfile of the NIOSH Registry of Toxic Effects of Chemical Substances*, 2d ed., ed. H. Christensen (NIOSH, Cincinnati, Ohio, 1976).

16. R.E. Kasperson, "The Dark Side of the Radioactive Waste Problem," in *Progress in Resource Management and Environmental Planning*, vol 2, ed. T. O'Riordan and K. Turner (John Wiley and Sons, New York, 1980), pp. 133-163.

17. U.S., Dept. of Transportation, *The National Highway Safety Needs Report* (DOT, Washington, 1976).

18. L.S. Robertson and P.L. Zador, "Driver Education and Fatal Crash Involvement of Teenaged Drivers," *American Journal of Public Health*, 68 (October, 1978), 959-965.

19. I. Burton, R.W. Kates, and G.F. White, *The Environment as Hazard* (Oxford University Press, New York, 1978).

20. *Perspectives on Benefit-Risk Decision Making* (National Academy of Engineering, Washington, 1972); G. Sinclair, P. Marstrand, and P. Newick, *Innovation and Human Risk* (Centre for the Study of Industrial Innovations, London, 1972); G. Calabresi, *The Costs of Accidents*, (Yale University Press, New Haven, 1970); N. Ashford, *Crisis in the Workplace: Occupational Disease and Injury* (Cambridge, MA., MIT Press, 1975); W.W. Lowrance, *Of Acceptable Risk* (Kaufman, Los Altos, California, 1976); Council for Science and Society, *The Acceptability of Risks* (Barry Rose for the Council, 1977); E.W. Lawless, *Technology and Social Shock* (Rutgers University Press, New Brunswick, New Jersey, 1977); *Managing Technological Hazard: Research Needs and Opportunities*, ed. R.W. Kates (Program on Technology, Environment and Man, Monograph 25; Institute of Behavioral Science, University of Colorado, Boulder, 1977); W.D. Rowe, *An Anatomy of Risk* (Wiley, New York, 1977); G. Calabresi, and P. Bobbitt, *Tragic Choices* (W.W. Norton, New York, 1978); R.W. Kates, *Risk Assessment of Environmental Hazard* (SCOPE 8; John Wiley for the Scientific Committee on Problems of the Environment, New York, 1978); M. Shapo, *A Nation of Guinea Pigs* (Collier Macmillan, New York, 1979); B. Fischhoff et al., *Approaches to Acceptable Risk: A Critical Guide* (NUREG/CR-1614; ORNL/Sub-7656/1, Oak Ridge National Laboratory, Oak Ridge,

Tennessee, 1980); *Energy Risk Management*, ed. G.T. Goodman and W.D. Rowe (Academic Press, New York, 1979); *Environmental Risk Assessment*, ed. A.V. Whyte and I. Burton (SCOPE 15; John Wiley for the Scientific Committee on Problems of the Environment, New York, 1980); National Council on Radiation Protection and Measurements, *Perceptions of Risk: Proceedings of the Fifteenth Annual Meeting...Held on March 14-15, 1979* (NCRP, Washington, 1980); *Societal Risk Assessment: How Safe is Safe Enough?*, ed. R.C. Schwing and W.A. Albers (Plenum Press, New York, 1980); *Society, Technology and Risk Assessment*, ed. J. Conrad (Academic Press, New York, 1980); U.S., Congress, House, Committee on Science and Technology, Subcommittee on Science, Research and Technology, *Comparative Risk Assessment: Hearings..., May 14,15, 1980*, 96th Cong., 2d sess., No. 129 (Government Printing Office, Washington, 1980).

Failures in Managing Technological Risk

Christoph Hohenemser, Jeanne X. Kasperson

2. Overview: Cases in Point

In this part, five authors probe and draw from particular cases of technological risk potential generalizations for improved risk management. Four of the five contributions are based on the symposium "Risk in the Technological Society," held 7 January 1980 at the annual meeting of the American Association for the Advancement of Science (AAAS) in San Francisco. The fifth is a more recent analysis that includes an essential complementary view.

Joseph Coates, the organizer of the AAAS symposium, leads off by arguing that present governmental structures are ill-suited to the task of risk management. Coates makes this point through a collection of delightful if sometimes depressing illustrations and anecdotes, which may well have lost something in the transition from his pungent oral delivery. Coates's prescription is radical: good risk management requires reform of the Constitution itself, which, he points out, was framed in the eighteenth century to deal with eighteenth-century problems.

Russell Peterson, President of the National Audubon Society, reflects on his experience as member of the President's Commission on the Accident at Three Mile Island. The accident, which occurred in March 1979, is by all accounts the nearest the U. S. or any other country has come to a catastrophic nuclear reactor failure. Peterson's outlook is skeptical if not pessimistic. Despite the widespread and vocal opposition to nuclear power in the United States, he is worried about the "mindset that nuclear power is safe, period," which still is widely held by risk managers both at the utilities and within the Nuclear Regulatory Commission. He warns that we pay too little attention to the risks of the full nuclear fuel cycle and worries that while developing nuclear power further, we shall short-change development of other energy sources.

In a companion piece on Three Mile Island, Stanley Gorinson, the legal counsel of the President's Commission, enumerates in detail instances that point to generic failures in the Nuclear Regulatory Commission, particularly in relation to issues raised by the Three Mile Island incident. Stated simply, Gorinson's point is that the NRC is currently too loosely organized and cannot effectively manage the safety of so complex and unforgiving technology as nuclear power. Gorinson identifies the present time as a crossroads and warns that reform of the NRC is a *sine qua non* if nuclear power is to have any future at all.

William J. Librizzi, formerly Deputy Director of the Hazardous Materials Division of the Environmental Protection Agency in New York State, reviews state and federal response to the still evolving situation at Love Canal, the nation's most notorious case of a chemical dump gone awry. Librizzi's review of necessity avoids certain sensitive issues bearing on current massive federal litigation against Hooker Chemical, the company that buried most of the wastes at the Love Canal site. His account details the government effort to cope with human exposure through monitoring, through on-site epidemiology, and through remedial action. The last includes plans, following the second of two Presidential emergency declarations, to evacuate the remaining 710 families of an original total of 949. Librizzi believes that the most important legacy of Love Canal concerns its implications for other hazardous waste sites. EPA estimates indicate that 80-90% of hazardous wastes are currently being disposed of in ways that will not meet the standards of recently passed laws.[1]

Closing out the group, Lois Ember, an associate editor of *Chemical and Engineering News*, focusses on the conflicts that have arisen in the past three years at Love Canal. One form of conflict concerns the interpretation of epidemiological data on Love Canal residents. Through a review of government and private studies Ember shows how time pressure, the residents' fears, and politics have confounded the proper interpretation of data. A third conflict centers on growing distrust between government agencies and the victims, who by Ember's account are sick with anxiety and anger and are unlikely to believe any assurance of safety in the future, no matter how long the studies are pursued. Ember offers few solutions, but she does leave the reader with much to ponder.

Taken together, the five papers offer a number of opportunities for comparison and further generalization. In our view, the most promising of these are:

● Conflict of interest pervades risk management in regulatory agencies.

● Public anxiety is a legitimate concern of risk management and cannot be treated as trivial or incidental.

● Science often falls short of regulatory needs and itself contributes to anxiety.

● Regulatory reform involves difficult choices.

Following is a brief discussion of each of these generalizations.

Conflict of interest. Conflict of interest pervades
the regulatory process: this much is explicit in four of
the papers and implicit in the fifth. Coates and Gorinson
both note how safety concerns at the NRC compete with, and
are at times overwhelmed by, the need to promote and protect
nuclear power technology. Peterson recalls how the President's Commission on Three Mile Island balked at including
a consultant who had publicly expressed doubts about the
safety of nuclear power. That nuclear power regulation is
thus flawed is perhaps an old story dating back to the dual
mission of the Atomic Energy Commission, an agency which
in the words of Coates, placed "blind optimists at the helm"
while retaining full responsibility for safety management.
What is surprising, however, is that, by Ember's account,
the Environmental Protection Agency betrays a similar conflict of interest. As a result of zealous efforts to score
legal points against Hooker Chemicals, half-baked, perhaps
uninterpretable epidemiological data have become public.
Instead of protecting public health (the overall mission of
the agency), the effect has been to induce additional
anxiety and to spur politicians to possibly premature
action toward mass evacuation. The EPA's conflict of interest is not unlike that of the police who conduct a high-
speed automobile chase through narrow streets in order to
apprehend a legal offender, and in their zeal to uphold the
law, seriously endanger the public they are charged to protect.

Public anxiety. Public anxiety is an inevitable product of risk management when the latter is highly visible
and comes in response to disaster. Confirmation of this
fact comes not only from the EPA's experience at Love Canal,
but also from the President's Commission on Three Mile
Island, which finds that the principal health effects of
the accident were anxiety and stress. Coates, Peterson,

Gorinson, Librizzi, and Ember all suggest in their own ways that risk managers must grapple with public anxiety and fear. Fear and anxiety may arise from lack of knowledge or from misunderstanding of uncertain science. Some fear and anxiety may be "real," "irreducible" and imbued with positive survival value. Yet to understand "why the people are scared" (as Coates puts it) does not solve the problem. An ethical response to public fear must consider the sources of fear and should not involve "fear management" by any means. This is a sensitive issue, and one for which society has worked out no definitive solutions. A tentative guideline might be that regulators should try to deal with fear arising out of lack of knowledge and the uncertainties of science while leaving untouched "real" or "irreducible" fears. The trouble with such a guideline is that it will be difficult to interpret and even more difficult to implement.

Uncertain science. A particular source of difficulty in risk management resides in the inherent uncertainties of science. Regulators and the public frequently demand definitive assessment of risk from science. Yet most scientific assessments are expressed in probabilistic language in which uncertainty is the rule, and victims are victims in a statistical sense. Both Three Mile Island and Love Canal involved the exposure of sizable populations to low levels of carcinogens, the effect of which is to produce some cancers in unknown or unknowable individuals. In each case regulators face a large number of people who feel themselves to be victims, even though we know from statistics that only a few will be actual victims.

In the case of Three Mile Island, the President's Commission minimizes statistical death by mouthing once again the cliche that "after many years of operation of nuclear power plants (there is) no evidence that any member of the general public has been hurt..."[2] This statement flies in the face of widespread knowledge that radiation is harmful, and that even under routine conditions, nuclear power plants emit radiation. The statement is literally true only if it is interpreted to mean that there are no data points (and hence no evidence) at the doses generally received by the general public. The statement is false and misleading whether one assumes linear[3] or linear-quadratic[4] extrapolation of human data from high dose to low dose. In either case, one must attribute a number of unknown and unknowable cancers and/or deaths to routine operation of nuclear power plants and to the release of radiation during the accident at Three Mile Island.

In the case of Love Canal, because there are no dose-response data of any kind for most of the 400+ chemicals that have been identified at the site, the situation is even more uncertain. As in the nuclear case, the victims are unknown and unknowable; unlike the nuclear case, no quantitative accounting of victims is possible, even in a statistical sense. And unlike the nuclear case, the experiments to determine the degree of harm are being performed on the exposed population itself.

Though Ember deals extensively with the problems of epidemiology and risk estimation at Love Canal, neither she nor the other four authors offer solutions to the problem of "statistical victims." The federal and state response at Love Canal has been to evacuate and relocate at government expense all *potential* victims. This suggests that one approach to dealing with statistical victims is to compensate all of them, regardless of whether they have experienced the consequences of exposure, such as cancer or other illness. Whereas such a principle appears to be feasible at Love Canal with 949 families, it becomes a pipedream at Three Mile Island, where much larger numbers are affected.

Regulatory reform. Four of the authors call for regulatory reform. Gorinson and Peterson discuss the need for greater attention to safety goals at the NRC and the banishment of a mindset that maintains that nuclear power is safe. Gorinson explicitly urges the separation of safety management from technology promotion. Ember argues for more and better science, more credible epidemiology, and greater attention to the fears of victims. Coates, in the most comprehensive statement on reform, calls for centralization and integration of risk management, with a high degree of the longitudinal coordination that prevails in large corporations.

On the face of it, the foregoing suggestions seem sound and defensible. Yet one wonders if a push toward more extensive epidemiology--destined to heighten uncertainty--at Love Canal is not in direct conflict with reducing anxiety; and whether integration of risk management does not exacerbate the conflict of interest that Gorinson and Peterson criticize. For if risk management is integrated as Coates wishes, does this not lead to internalized conflict between risk and benefit, much as in the defunct AEC or in the boardroom of the large corporation? We have no answers of our own for these dilemmas and take note of them here only to suggest that regulatory reform is not a simple issue. It may, instead, involve a choice between an adversarial system, beset by serious and open conflict and a high level of anxiety (the present U. S. system), and a system with internalized conflict and perhaps a much lower level of anxiety (approximated by Britain and France). Neither system of risk management seems ideal in itself, and each leaves open significant mechanisms for mishandling technological risk.

References and Notes

1. U.S., President's Commission on the Accident at Three Mile Island, *The Need for Change: The Legacy of TMI* (Government Printing Office, Washington, 1979).

2. Fred C. Hart Associates, *Preliminary Assessment of Clean-up for National Hazardous Waste Problems* (Environmental Protection Agency, Washington, 23 February 1979), p. 25.

3. National Research Council, Advisory Committee on the Biological Effects of Ionizing Radiations, *The Effects on Populations of Exposure to Low Levels of Ionizing Radiation: Report* (National Academy of Sciences, Washington, 1972).

4. National Research Council, Committee on the Biological Effects of Ionizing Radiations, *The Effects on Populations of Exposure to Low Levels of Ionizing Radiation: 1980* (National Academy Press, Washington, 1980).

3. Why Government Must Make a Mess of Technological Risk Management

Introduction

An alternative title for this discussion might well be:
Why the People Are Scared. For, as other papers (Chapters
4-7) in this volume demonstrate, ineptitude in risk management has fostered fear. "Love Canal" and "Three Mile Island"
have become scare words that symbolize our mismanaged technological society.

At the outset, let us ask why risk is today a new or
newly important subject for American citizens and American
government. My general answer is that citizens and government have failed to understand many aspects of our technoeconomic system, particularly the fact that America is today
the most technologically intensive society that has ever
existed. This paper's discussion centers on this general
hypothesis.

Risk management must deal with conflict, for conflict
is inherent to the problem. The issue of freon in the atmosphere, for example, is far more than a research problem for
scientists; it involves a struggle between highly important
short-range economic interests and unresolved doubts about
harmful effects on future generations.[1] Similarly, in the
case of motorcycle helmets, which many cyclists shun, it is
necessary to balance the absolutely parochial interests of
the individual against the costs to society of maintaining
him/her following a nonfatal accident.

According to our laws and traditions, in some cases conflict about risk can be resolved by individuals. When
individual and general well-being are in conflict, however,
we have turned most often to government, including the

bureaucracy, the Congress, state legislatures, and the courts. The remainder of this paper explores the source of risk, how risks are handled by corporations and individuals, and why government as presently constituted must make a mess of resolving the conflicts inherent in risk management.

The Roots of Risk in the Technological Society

Most risks derive from technology. In his book, <u>Technology and Social Shock</u>, Lawless[2] of the Midwest Research Institute found by a fairly straightforward tally that since World War II there has been a steady flow of instances of technological failure or risk of sufficient note to make the national press. A significant proportion of these involved gaffes, blunders, or failures in hazards management. Indeed, based on the Lawless study, I estimate there has been approximately one significant technological blunder each week, week in and week out, for thirty-five years. This alone leads one to the conclusion that there is something amiss in the social management of technology.

My favorite case is diethylstilbestrol (DES), a versatile drug which both inhibits and promotes conception, and still enjoys use as a "morning after" pill, albeit against a label-warning to the contrary.[3] DES was first tried on chickens to caponize them; it was later tried on cattle to make beef more tender; and finally it was used on pregnant women to prevent miscarriage. Problems emerged in all three uses. In particular, as far back as 1968 it was shown that DES causes cancer in daughters of women who used DES as a morning-after pill.[4] One wonders what is wrong with a system that needs to experiment on three species, all of which are important to us, before concluding that there is something not entirely safe about DES.

This case, and the thousand others like it, flow from the fact that Americans today occupy a most unusual situation in world history. For the first time, and to an extent greater than anywhere else in the world, we live in a totally human-made world. There is little that one encounters which is not the product of human enterprise. You cannot even get away for a vacation without Smokey the Bear or Coast Guard technology watching over you. We have built this technological world in about seven decades, and we do not know much about its elasticity, its forgivingness, its healing power, or its capability to recover. And yet we act as if we do. We have spawned a fundamental mismatch between our technology and institutions established by a simpler, more direct, more nature-driven society.[5]

A nice symbol of this mismatch is a game played by British university dropouts. The point of the game was to go into one of the phone booths that are clustered on London streets and call yourself in the next booth, without paying. The game had a charming anti-technological twist: players tried to reach the booth next door by using the maximum number of linkages, for example, by dialing from London to Paris, to Milan, to Stockholm, to Bern, and so on around the world. The British Postal Service finally apprehended the pranksters and awarded them amnesty, but under one condition: "Tell us how you did it."

The telephone game is not only symbolic of mismatched technology and institutions but a paradigm of many technological systems. The people who build, design, plan, execute, sell and maintain complex systems do not know how they <u>may</u> operate. If and when the first half-billion dollar theft takes place in the growing system of electronic funds transfers, it will be for this reason; for that system is beyond the understanding of its designers.[6]

One of the best documented failures in anticipating the full effect of technological systems dates back 44 years. At that time, the Army Corps of Engineers initiated a major national program to control waterways. Current evidence from the academic community indicates that many of the Corps' civil works (levees, dams, channels) have worsened the problem in two distinct ways.[7] They made channels narrower, water speed faster, and thus produced a rate of flooding higher than in the past. Even more significant than this negative hydrologic side of flood control, however, is the fact that the Corps of Engineers programs have instilled a kind of complacency which reenforces the greed of the average local realtor and the myopia of the average local official. We are building furiously on the floodplains nominally "protected" by flood control programs. Thus, instead of "small" investments at a given site being destroyed every twenty years, the Corps programs insure major disasters at more infrequent intervals, with societal costs that actually exceed those prevalent during the pre-control era. By way of example, a prime candidate for flood disaster is Orange County, California. That flood will produce damage involving several billions of dollars and will affect 10,000 to 15,000 people. It may take 75 years or it may take 50 to occur; but it will occur in the lifetime of many readers of this volume.

Given the rate of propagation in our human-made world, disasters now occur at an unprecedented speed and on an expanding scale. The disaster of thalidomide resulted wholly

from a failure to test the new drug in the usual way of simpler and more parochial times. Fully operating, legitimate
market forces spread thalidomide to a wide population. Similar market forces were responsible for the rapid deployment
of automobile manufacturing defects, which in one case neces-
sitated the recall of 320,000 General Motors vehicles.

Some of the risks from technology arise from simple ignorance or the failure to be aware of the potential danger.
People who are not cognizant of the risk may act as if they
were not accountable for internalizing social values in their
technological operations. The case of Kepone contamination
in Hopewell, Virginia, may be a case in point.[8] An ordinary
run-of-the-mill, routine, chemical engineering operation led
to a contamination problem that according to the Corps of
Engineers will cost $1 to $7 billion dollars to clean up.[9]
Specific effects include the closing of the James River to
commercial and private fishing, the hospitalization of numerous factory workers, and widespread anxiety in a broader
community. The Love Canal situation described elsewhere in
this volume (Chapters 6 and 7) has a similar structure; here
too, a sequence of decisions made by the ill-informed led to
a situation in which several hundred people are living on top
of a chemical dump that continues to emit significant quantities of toxic vapors and liquids. The linkage between the
knowledge system and the operating, managerial system was
weak.

One wonders how many "routine disasters" we can tolerate
before we realize that there is something intrinsically wrong
with the way American society operates. At the same time, it
is important to realize that the problem is not just an American, but a world-wide problem. The Kepone contamination in
Virginia has its counterpart in the dioxin clouds over
Seveso, Italy, and in Minamata disease in Japan. The international scene reminds us that there are helpless victims of
technological mismanagement everywhere. Thus, the Swiss
built in Seveso, Italy, a plant that would not have passed
Swiss rules and regulations and thereby victimized an unaware,
unprotected rural Italian populace.[10] Similarly, the Japanese, after experiencing severe mercury poisoning at Minamata,
exported to Thailand the causative chlor-alkali technology
which no longer meets Japanese safety criteria.[11] In short,
there are victims everywhere who willingly or unwillingly
accept the short-term economic reward of rapid industrialization in return for longer-term risks that they cannot begin
to understand.[12] Unfortunately, there also are those ready
to exploit the innocent potential victims.

Technological Risks as Side Effects

A major factor in generating risks is that new knowledge intrinsically, not accidentally, generates uncertainty and ignorance, which necessarily leads to unanticipated side effects that we characterize as risks. The more important the new knowledge, the greater the degree of ignorance that it creates. Though many of us know this, we rarely, as a society, integrate this basic observation into our thinking and planning.

A minor example of the side-effects problem arises in scuba diving, an increasingly popular form of recreation. There are in the United States about 16,000 women between the ages of 16 and 40 who scuba dive. They are likely to be bright, hardworking, athletic, no-nonsense women. It was discovered a few months ago that fetuses can suffer the bends and are more susceptible than their mothers--an unexpected, little side effect. The new knowledge which permitted the the invention of scuba diving and its commercialization now creates a problem that was unthinkable 25 years ago. It is a rather minor problem in some sense, unless you happen to be one of what I estimate to be four unborn children who develop fetal bends.

The side effects of technology are more serious when viewed in the context of corporate actions, which to a large extent take place in a so-called free market. For corporations, it is intrinsic to their mode of operation to "externalize" all the costs they can. Externalizing costs implies avoidance of responsibility for side effects; or, alternatively, passing the responsibility on down the chain of distribution to the customer, to the public at large, or to government. Only informed individual and governmental action can force the internalization of the externalities. The pollution by the automobile of every major city in the United States is an example of externalization, as are thalidomide, Seveso, and Minamata disease. Except for the occasional idealist among corporate managers, there is in the present corporate structure nothing that acts as an effective internal force to prevent the dumping of the hazardous wastes of every manufacturing process.

Love Canal, described by Librizzi and Ember (Chapters 6 and 7) is an example of the private sector's operating at "its very best" by externalizing that which cannot be sold, marketed, or given away. By my calculations, preventing the Love Canals of the United States would have cost less than a cent a pound for all the organic chemicals produced since

World War II. But we need some new calculus which permits
society to ask: "Do we want to pay that cent in the form of
Love Canals, or do we want to pay for it directly, as a minor
increase in the price of bug spray and garbage bags?" Every
nylon rug, every polyester shirt, half of the things one can
see in an average room or office contribute to Love Canals
because of the way that the corporate sector externalizes its
costs. It is as American as apple pie and as univeral as the
principles of Adam Smith.

   If side effects are difficult for the corporate sector
to deal with, they are even more of a problem for government,
particularly the bureaucracy. For government and bureaucrats
are not subtle. Even if motivated by legislation or the
courts to look for side effects, bureaucrats do not generally
know how, because they lack the experience, tools, and motivation for thinking in systems terms. In the few cases where
government officials understand what to do, the structure of
regulatory institutions often stymies effective action.

   An example of the misguided institutional structure of
government is the way that nuclear power has been handled.
As Harold Green has pointed out repeatedly,[13] the Atomic
Energy Commission (AEC) and its descendants have had the dual
responsibility of regulation and advocacy. Being basically
technologically oriented, the advocates shaped and framed the
regulations to facilitate the development of atomic energy.
Over a period of twenty years, the AEC's commissioners were
not suitably mindful of the fact that they were not dealing
with a usual technology; nuclear power did not evolve the way
other technologies have, via a path that is initially small-
scale, trial-and-error, and subject to effective feedback.
Instead, nuclear power was developed in a hothouse that was
almost the antithesis of every successful technology in the
United States. When you put advocates in charge of regulating such a technology, when you put blind optimists at the
helm, you drive out reason, commonsense and prudence. Nor
has any of this changed with the division of the old AEC into the Department of Energy and the Nuclear Regulatory Commission. Elsewhere in this volume, both Russell Peterson
(Chapter 4) and Stanley Gorinson (Chapter 5) in their respective discussions of the accident at Three Mile Island
recall the NRC "mindset" that is still fundamentally geared
to nurturing a growing industry. A similar point has been
made by the Kemeny Commission[14] on which Peterson served and
for which Gorinson was legal counsel.

   If side effects of technology produce the risks we fear,
what drives the continuing evolution of technology? Here I
want to make just two elementary points and then leave the

rest to your imagination. The first is that virtually no
technoeconomic enterprise in the United States has historically had to deal with anything but minimal initial acceptance
criteria: (a) can you do it, will it work? (b) will it sell,
or will government subsidize it? and (c) by some commonsense
standard, is it safe to use? These simple criteria which
built America ignore side effects and hence the majority of
problems of risk. My second point is that there is a mismatch between traditional technoeconomic planning criteria
and what ultimately becomes important. The three simple
criteria for initial acceptance simply do not suggest the
real consequences that ultimately evolve in our complex
world. This goes beyond side effects and to the very structure of our society. Could anyone have imagined or predicted
that the automobile would not only provide individual transportation but also completely restructure our use of space,
in particular the layout of our cities? To begin with, technology is almost always introduced on the basis of substitution, as a new thing which does an old job better. Yet the
new and the old are never truly congruent. While we pay
attention at the outset to the overlapping parts of the substitution, what dominate in the long run are the unsuspected
aspects of substitution resulting from those features of the
new that are different from the old.

In sum, therefore, risks arise from side effects of
technology that lie outside the purview of the usual economic
forces or institutional structure; technology itself is
driven "forward" by short-term forces of substitution which
almost never accurately reflect the ultimate changes that
technology will bring.

The Social Aspects of Risk and Risk Management

Several noteworthy social factors tend to interact with
our handling of risk. As members of a middle-class society
that has been 80% center-cut middle class for perhaps three
or four decades, Americans tend to institutionalize their
problems. This comes home to roost in things like the environmental impact statement, OSHA regulations, and scores
of related controls. As members of the middle class we desire institutional intervention where we think it will protect us, but we deplore such intervention when it hits the
sector where we earn our living. The problem is to strike
a balance. A related aspect of institutionalizing the problem of risk is a deep-seated confusion in middle-class Americans between symbol and action, between wish and accomplishment. All too often we start out to solve a problem by making a gesture and then fail to watch what actually happens.

For example, has anyone asked what has been the real effect of all those reams of paper printed with environmental impact statements--at a cost of perhaps half a billion dollars?

A deep current underlying the middle-class view of risks is ignorance, particularly as to the workings of technology. Indeed, there has probably never been a greater wave of ignorance about the human environment than that which is now washing over the middle class. Ask the ordinary citizen a series of simple questions about the technological world:

What is nylon? Where does your sewage go when you flush the commode? What is polyester? Why do fluorescent bulbs flicker? Where does the picture go when you turn off the TV set? What exactly is a rocket? Where does oil come from?

Then examine the answers, and ask yourself how such a society will handle technological risks. In a middle-class society that is ignorant about its technical workings, when trouble arises there are only a very few things that can happen. Most often we will summon the priest for exorcism or call upon the political leader to produce a scapegoat.

The problems of the middle class, however, go well beyond that. The middle class tends to be both risk aversive and risk embracing; it is risk aversive collectively and risk embracing individually. It wants to go hang gliding and skiing--two of the things closest to madness created by technology--yet it will not tolerate any DC-10s falling into the ocean. Before we can hope to build reasonable regulatory institutions we have to do something about this schizoid view, or at least begin to understand the roots of it. Such a beginning has been made by the psychologists Slovic, Fischhoff, and Lichtenstein, in their article on perceived risk (Chapter 10).

Government Role in Risk Management

In a middle-class society that institutionalizes its problems, it falls to government to decide how to manage many of the risks that corporations would externalize and that individuals do not understand. To evaluate the role of government in risk management, do not consider all that government might do, i.e., its maximum measures. Consider only the minimum roles that government ought to play in assisting us as citizens, business leaders, and professionals in coping with risks. Three minimum roles of government are:

• to supply or assure a supply of adequate

information with regard to the risks under
discussion;

- to assure that there are choices which are
open and available in a timely fashion;

- to orchestrate the means at the government's
disposal for coping with risk.

Government as a risk manager fails on all three counts.

Our society, our complex human-made world, has as its
dominant institution the bureaucracy, not by accident, but
simply because there is no choice. Complexity demands expertise. The expert must be housed somewhere. That place is
called a bureaucracy, that is, an organization with a hier-
archy of functions, an internal reporting system, and a division of labor. Intrinsic to the utilization of expert knowledge is this division of labor. Be it the filling out of
form 7730 or the construction of scintillation counters, it
is impossible to escape bureaucracy and its experts.

Yet from the work of Max Weber and a dozen others following him in the last fifty years come a number of melan-
choly and alarming propositions about bureaucracies.[15]
Bureaucracies are conservative; they exist not to serve the
public interest but to preserve themselves. They lie and
shirk responsibility where necessary; they shun controversy.
Yet, in our world fraught with problems and bountiful in
opportunities, bureaucracies are our dominant institutions
for handling the risks which other sectors (also bureaucracies) of society have bequeathed to them.

There is abroad, with regard to risk, an interesting
expert-bureaucrat game called "the search for acceptable
risks." This game serves only one bureaucratic purpose, to
create the illusion of doing something while safely accomplishing little or nothing. Only a very small fraction of
risks can be set by anything like a defensible standard.
The long-term, major risk levels in society are the outcome
of an evolutionary process of trial and error conducted by
the affected parties. Yet it is in the interest of bureaucrats to disburse money, and it is in the interest of experts
to accept it to pursue their mad, mad search for standards.
That search is more convenient than an explicit recognition
of the traditional American method of dealing with the prob-
lem--true choice counterpointed by an approach that is in-
cremental, experimental, and trial-and-error to improve,
reinvent, and redesign processes and devices. Those two
mentalities are as antithetical to each other as a schizoid

society will permit. For those interested, the search for
quantitative risk acceptance criteria is well represented in
this volume, in papers by Deisler (Chapter 15), Okrent (Chapter 13), and Starr and Whipple (Chapter 14). Quantitation
should be the guide to choice and not the choice itself.

If government's search for workable risk management is
not defeated by the bureaucratic mentality and a vain search
for "acceptable risk," then surely it faces the greater
threat that its own structure will permit or produce increasingly massive failures of technology. The governmental
structure in the United States is in a dreadful state for
managing technology because the fundamental legal instrument,
the Constitution, is at odds with the technological structure
of our society. There is no agency of government that has
even the minimum responsibility for collecting, organizing,
and disseminating information on any given subject. Nor
does any single agency have responsibility for control, regulation, or policy. The mismatch between the risks to be controlled and the structure that we expect to do the job comes
in three layers--federal, state, and local. Each layer is
structured into an inchoate system of departments and
agencies. By my estimate, there is in America no significant
technology that is not the province of at least a dozen government agencies. Contrast this picture with that of corporate giants that now extend across the international boundaries. There is no Fortune 500 or Fortune 1000 company that
is not toally integrated on at least a national scale.

To see why there is a relationship between our management of technology and the Constitution, consider the case
of the amendment that guarantees freedom of the press. If
you reflect on it, freedom of the press has not always been
with us but is directly driven by technology. There is no
need to guarantee freedom of the manuscript simply because
it is not worth trying to constrain manuscripts--the risks
to the established order are self-limiting. Society has
always enjoyed freedom of the manuscript, and this freedom
is enjoyed even in the Soviet Union today. Anything that is
limited to a single copy is permissible. How far can it go?
Clearly, the circulation of a manuscript is self-limiting.
It is only with inventions such as cheap paper, the rotary
press, and movable type that one creates a crisis for government and in turn engenders its desire to censor. In
response to this new pressure from government a new freedom--
freedom of the press--is created. In my estimation, at least
half of the provisions of the Bill of Rights reflect responses to the technological issues of the 18th century. We
have not brought the Constitution up-to-date to our present
world, much less to the world of the third millennium.[16]

Still other problems of American government, derivative of its constitutional and bureaucratic structure, stem from the fact that it is centrist, or middle-of-the-road. The political system in the United States entertains no choice, no diversity beyond Tweedledum and Tweedledee, because the political process drives out all radicalism and most substantial reform. The only way to survive for thirty years in the political system is to spread a vague umbrella of nothingness over the largest possible group of constituents. The present political system is one in which there is virtually no member of Congress who is not motivated more by a dread of being unseated than by any desire for real leadership.

Conclusion

I have shown through examples that the risks we face arise as unanticipated side effects of technology; that technology itself is driven by institutional forces that fail to anticipate side effects, let alone the ultimate changes that will take place in society. I have suggested that the corporate structure and the "free market" work together to create risks in an intrinsic, not accidental manner. In this situation individual action in a limited sense and government action in a broader sense are the only means at society's disposal for protecting us from the risks of technology.

But government is poorly suited to its task due to bureaucratic blindness and acts of self-preservation. This is because of antiquated constitutional arrangements and because our centrist politics drive out most needed reforms. It therefore becomes inevitable, as suggested in my title, that government under present circumstances <u>must</u> make a mess of risk management.

If we are to avoid the failures of government in risk management, we must achieve at least three minimal performance standards, as detailed earlier: government must assure that there is adequate information, government must provide for adequate choices in a timely fashion, and government must orchestrate the means for coping with risk. And if these means are, as I have suggested, fundamentally inadequate on constitutional grounds, then I see no way of avoiding the restructuring of the institutions that currently are responsible for government risk management.

Key elements which will be part of such a restructuring are:

- <u>Forecasting</u>. At present it is difficult to point to a single, interesting, solid, well-rounded

forecast of anything important from either government or the private sector. We need forecasting
so that the rapid development of technology does
not outpace our identification and recognition of
risks.

- Flexibility. At present it is almost antithetical to the notion of writing laws and regulations
  that they be flexible. Yet a rapidly changing
  technology and a continuously evolving future
  make very strong demands for flexibility in dealing with risks.

- Feedback. Though we have feedback in our current
  political process, it is available only in the
  slowest and most obsolescent way. Effective and
  more rapid feedback is essential for future risk
  management and will require that there be specific institutions that provide it in a timely way.

References and Notes

1. National Research Council, Assembly of Life Sciences,
    Freon Question: Propellant in Aerosol Spray Cans
    (National Academy of Sciences, Washington, 1974).

2. E. W. Lawless, *Technology and Social Shock* (Rutgers
    University Press, New Brunswick, N.J., 1977). This book
    comprises some 45 case histories, drawn from a pool of
    approximately 300 technological hazards. It was clear
    from discussion with the author that carried further,
    the method of selection would have yielded perhaps a
    thousand instances. The criteria for what constitutes
    a technological gaffe is an elastic concept. Notice in
    the national press is one sensible criterion. What is
    less important than the actual number is the consistency
    and steady flow over three decades of these major and
    minor failures of technology.

3. The DES case is discussed in interesting and informative
    brevity by Lawless, *Technology and Social Shock*. For a
    more detailed discussion see M. S. Shapo, *A Nation of
    Guinea Pigs* (Collier Macmillan, New York, 1979), pp.
    163-190.

4. N. P. Napalkov and V. A. Alexandrov, "On the Effects of
    Blastomogenic Substances of the Organism During Embryo-
    genesis," *Zeitschrift für Krebsforschung and Klinische
    Onkologie/Cancer Research and Clinical Oncology,* 71
    (1968), 32-50.

5. J. F. Coates, "Technological Change and Future Growth: Issues and Opportunities," in U. S. Congress, Joint Economic Committee, *Technological Change, U. S. Economic Growth for 1976 to 1986: Prospects, Problems and Patterns,* (Government Printing Office, Washington, 3 January 1977), vol. 9.

6. Arthur D. Little Inc., *The Consequences of Electronic Funds Transfer* (Arthur D. Little, Inc., Cambridge, MA., 1975), pp. 165-66.

7. A solid but journalistic account of the short-term benefits and long-term risks engendered by U. S. Corps of Engineers civil works projects is presented by Wesley Marx in *Acts of God, Acts of Man* (Coward McCann and Geohagen, New York, 1977).

8. National Research Council, Panel on Kepone/Mirex/Hexachloropentadiene, *Kepone/Mirex/Hexachloropentadiene: An Environmental Assessment* (National Academy of Sciences, Washington, 1978); B. E. Suta, *Human Population Exposures to Mirex and Kepone* (EPA 600/2-78-045; U. S. Environmental Protection Agency, Washington, 1978).

9. The Corps of Engineers studies and related materials are presented in U. S. Environmental Protection Agency, Office of Water and Hazardous Materials, Criteria and Standards Division, *Kepone Mitigation Feasibility Project: Kepone-Contaminated Hopewell/James River Areas* (EPA/440/5-78/004, Washington, 9 June 1978). See also B. McAlister, "Kepone Removal May Cost Billions, If It Is Possible," *Washington Post,* 8 February 1978, p. 9; M. H. Zim, "Allied Chemical's $20-Million Dollar Ordeal with Kepone," *Fortune,* 98 (11 September 1978), 82-91.

10. J. G. Fuller, *The Poison That Fell From The Sky* (Random House, New York, 1977); P. Lagadec, *Developpement, environnement, et politique vis-à-vis du risque; la cas de Italie: Seveso* (Laboratoire d'économetrie, Ecole Polytechnique, Paris, 1979).

11. Minamata disease is discussed by Lawless (*supra,* note 2) as part of a general discussion of "Mercury Discharges by Industry," pp. 248-59. Export of Japanese mercury polluting industries to Thailand has been described by S. Suckcharoen, P. Nuorteva, and E. Häsänen, "Alarming Signs of Mercury Pollution in a Freshwater Area of Thailand," *Ambio,* 7 no. 3 (1978), 113-116.

12. An editorial in *Nature,* 273 (1978), 415, notes several

instances of transfer of dangerous industrial facilities from the industrialized nations to the Third World, including the growth of "...the benzidine dye industry in India, the shift of asbestos textile production from the United States to Latin America, and plans to establish vinyl chloride and arsenic production facilities in South Korea and the Philippines, respectively."

13. H. Green and A. Rosenthal, *Government of the Atom* (Atherton Press, New York, 1967).

14. U.S., President's Commission on the Accident at Three Mile Island, *The Need for Change: The Legacy of TMI* (Government Printing Office, Washington, 1979).

15. Anthony Downs has summarized the state of knowledge with regard to the rules governing bureaucracy in some 315 hypotheses and subhypotheses in his classic book, *Inside Bureaucracy* (Little Brown, Boston, 1967). A useful guide to the literature is M. K. Mohapatra and D. R. Hager, *Studies of Public Bureaucracy: A Selected Cross-National Bibliography,* Exchange Bibliography 1385, 1386, and 1387 (Council of Planning Librarians, Monticello, Illinois, 1977).

16. R. G. Tugwell has most diligently pursued the need to rewrite the American Constitution. His *The Emergent Constitution* (Harpers, New York, 1974) is his last major work on the subject. The need for a new constitution is treated briefly and incisively in his *A Model for a New Constitution* (Center for the Study of Democratic Institutions, Santa Barbara, California, 1970). An alternative and exciting treatment of the same subject is Leland D. Baldwin's *Reframing the Constitution: An Imperative for Modern America,* (ABC Clio, Santa Barbara, California, 1972).

Russell W. Peterson

4. Three Mile Island: Lessons for America

The accident at the Three Mile Island Nuclear Power plant offered us many lessons, but it is not clear that we have learned many of them. In this paper I shall review these lessons from the perspective that I gained as a member of the President's Commission on the Accident at Three Mile Island, also known as the Kemeny Commission. The Commission was chaired by John Kemeny, the President of Dartmouth, and consisted of a total of twelve individuals representing a number of interests. It was the charge of the Commission to "conduct a comprehensive study and investigation of the recent accident involving the nuclear power facility on Three Mile Island in Pennsylvania."[1]

Findings of the Kemeny Commission

It is generally agreed that the radiation released as a result of the accident was minor, and except for severe, but short-lived mental stress, the health effects of the accident will probably be so small that it will be impossible to detect them. In contrast, physical damage to the plant was devastating. Costs approaching $2 billion are likely to be incurred over the four years required to bring the damaged TMI-2 plant back into operation. This is more than double the original cost of construction, and this figure is so high largely because of the massive clean-up that is required. In addition, TMI-1, the undamaged plant directly adjoining TMI-2, has also been kept out of operation since the accident.[1]

The accident is attributable to a number of factors, but primarily to human failure to understand fully the complex signals being given by instruments in the control room in the course of the accident. Let me quote from the Kemeny Commission report:[1]

We are convinced that if the only problems were
equipment problems, this Presidential Commission
would never have been created. The equipment was
sufficiently good that, except for human failures, the major accident at Three Mile Island
would have been a minor incident. But wherever
we looked, we found problems with the human beings
who operate the plants, with the management that
runs the key operation, and with the agency that
is charged with assuring the safety of nuclear
power plants.

In the testimony we received, one word occurred
over and over again. That word is "mindset."

After many years of operation of nuclear power
plants, with no evidence that any member of the
general public has been hurt, the belief that
nuclear power plants are sufficiently safe grew
into a conviction. One must recognize this to
understand why many key steps that could have prevented the accident at Three Mile Island were not
taken. The Commission is convinced that this attitude must be changed to one that says nuclear power
is by its very nature potentially dangerous, and,
therefore, one must continually question whether
the safeguards already in place are sufficient to
prevent major accidents. A comprehensive system
is required in which equipment and human beings
are treated with equal importance.

The Presidential Commission's principal conclusion was as
follows:

To prevent nuclear accidents as serious as Three
Mile Island, fundamental changes will be necessary
in the organization, procedures, and practices,
and above all, in the attitudes of the Nuclear
Regulatory Commission; and to the extent that the
institutions we investigated are typical, of the
nuclear industry.

The report further states:

We do not claim that our proposed recommendations
are sufficient to assure the safety of nuclear
power. Our findings simply state that if the
country wishes, for larger reasons, to confront
the risks that are inherently associated with

nuclear power, fundamental changes are necessary if
those risks are to be kept within tolerable limits.

It appears to follow, then, that until the necessary
fundamental changes are made, the risks of nuclear power are
intolerable.

Continuing with a quote from the report:

While throughout this entire document we emphasize
that fundamental changes are necessary to prevent
accidents as serious as TMI, we must not assume
that an accident of this or greater seriousness
cannot happen again, even if the changes we recommend are made. Therefore, in addition to doing
everything to prevent such accidents, we must be
fully prepared to minimize the potential impact
of such an accident on public health and safety,
should one occur in the future. The fact that
too many individuals and organizations were not
aware of the dimensions of serious accidents at
nuclear power plants accounts for a great deal of
the lack of preparedness and the poor quality of
the response.

In response to this concern, the Commission unanimously voted
that

the NRC or its successor should, on a case by case
basis, before issuing a new construction permit or
operating license...condition licensing upon review
and approval of the state and local emergency plans.

An even more drastic condemnation of nuclear energy
development than that of the President's Commission has just
been made by the special inquiry group established by the
Nuclear Regulatory Commission (NRC) to investigate the TMI
accident.[2] Its views closely parallel those of at least
four of the members of the President's Commission.

It is important to recognize that both groups strongly
recommend that plans for evacuating residents within many
miles of nuclear plants be developed before a plant is
authorized to operate. Couple this with the sworn testimony
of nuclear physicists in top positions at the NRC: in the
event of a future nuclear plant accident, given the same
type of information obtained from TMI early in the incident,
they would recommend evacuation as they did in the case of
TMI. Together these facts demonstrate that when individuals

have the opportunity to dig deeply into the operation of nuclear power plants, they conclude that it is only a matter of time before an accident releasing a devastating amount of radioactive material occurs. Why else would we need evacuation plans?

The Chance of Catastrophic Accident and Public Fear

The rapidly increasing number of nuclear power plants around the world increases the chance that a catastrophic accident will occur someplace, sometime in the next few years. If and when it does occur, according to a 1965 Atomic Energy Commission report,[3] it <u>could</u> kill 45,000 people, injure 100,000 others, and require land use restrictions for hundreds of years in an area the size of Pennsylvania. In the NRC's 1975 Reactor Safety Study (Rasmussen Report), it was noted that the calculated probability for such large consequences is quite low, and that less severe consequences are more likely.[4] But because the 1975 estimates have been widely challenged and debated, they have done little to relieve public anxiety.

The concrete lesson of TMI is in fact that whatever the probability estimates are, nuclear catastrophe <u>can</u> happen. Public anxiety and fear are in this sense justified. At TMI the whole world became intensely concerned about a "near miss." TMI became one of the most highly covered news events in history. Heads of state, government agencies, launched major investigations. President Carter visited the plant during the accident. Plans for referenda in other countries to restrict nuclear energy were directly impacted. Directly after the accident a Harris Poll in the United States showed an 11% increase (over October, 1978) in the number of respondents opposed to the construction of nuclear power plants.[5] All this occurred in a climate in which the public, even prior to the accident, rated nuclear power as having the highest overall risk of any technology (Slovic et al., Chapter 10).

Besides the direct problems for affected individuals, the fear of nuclear power creates a major problem for our society. Can you imagine how people will react when we <u>do</u> have a major release of radioactivity? They will, in my opinion, almost certainly demand that we put an end to the era of nuclear energy. If in the interim we have tied our way of life to nuclear electricity, the economic and social costs will be major indeed.

Certainly we do not want to base our plans for the future on unwarranted fear. If people are unduly concerned

about nuclear risks because of lack of information or the
wrong kind of information, then we should strive to reduce
that concern by adequately informing them of the nature of
the risk. On the other hand, if the risk is large and real,
then to belittle the concern that people have is wrong
indeed. Then the action should be to remove the risk.

The fear of people is hardly alleviated by the fact that
the nuclear industry and the insurance companies consider
the risk so high that they are unwilling to assume the financial liability for nuclear accidents. As a result, the federal government has had to limit, through the Price-Anderson
Act, the liability of the industry in order to make nuclear
power viable. The effect of the federal limit is that nuclear power is insured by the concerted action of all tax-
payers. The need for the Price-Anderson limit and the estimated $2-billion cleanup and reconstruction cost for TMI-2
make the approximate $100-billion investment in nuclear
plants now under construction a dubious economic proposition
indeed!

The Cumulative Impact of Nuclear Power

Although the potential for a catastrophic accident
raises our prime concern, a more serious risk over the long
run could come from the cumulative impact of countless minor
releases of radioactivity. A little release here, a little
there, so what? That is what we said about toxic chemicals
in the past, and now, for example, Americans, Europeans,
Israelis, Japanese, and even the penguins in Antarctica have
life-deforming PCBs in their tissues.[6] That is what we said
about emissions from coal in the past, and now these lead to
acid rainfall, hundreds of miles from its source, sterilizing
lakes and stunting forest growth. That is what we said
about cutting a few trees in the vast Amazon River Basin[7,8]
and now it looks as though they will be gone in another 30
years, along with one million tropical species of plants
and animals.

It is time to concern ourselves with the cumulative
impacts of nuclear energy. Fortunately, the world is still
in the early stages of this development, and we are not yet
addicted to it. But accumulating threats to life from this
fledgling industry are all around us:[9-12]

(1) The widespread scattering of lethal uranium
tailings.

(2) The radioactive waste accumulation from the ore
processing and enrichment plants.

(3) The in-house, highly radioactive waste dumps of the operating plants.

(4) The trucks hauling extremely hazardous fuel and waste.

(5) The government dumps reluctantly receiving large quantities of contaminated clothing and equipment.

(6) The minor releases of radioactivity, accidental or intentional, from operating plants.

(7) The Three Mile Island accident, and the ongoing effort to clean up the huge quantities of lethal fission products that were released into the buildings and cooling water during the accident.

(8) The abandonment of the highly contaminated nuclear fuel reprocessing plant at West Valley, New York, which could be described as one of our accumulating islands of toxicity.

(9) The use of temporary employees as "sponges" to carry out brief tasks in highly contaminated areas to save regular employees from excessive exposure.

(10) The threats from radioactive releases that might be caused by sabotage and terrorism.

Each new nuclear plant adds to these problems. This is especially true in the developing world where the technical infrastructure is less adequate to cope with the complex, dangerous processes involved.

At the same time, outside the United States, the world is moving toward an even more dangerous process, the breeder reactor. Instead of water for a coolant, the breeder uses molten sodium which can react explosively with water. The attractiveness of the breeder lies in its capacity to produce plutonium, a nuclear fuel, in quantities greater than the fuel consumed. Eventually, when rare fissionable uranium runs out, the breeder could be used to replace the present generation of uranium-burning reactors. But for plutonium to be used as fuel it must first be separated from the fuel elements in which it is produced. This requires reprocessing of very large quantities of radioactive materials. Plutonium presents, in addition, the hazard of diversion

for weapons purposes; fifteen pounds is sufficient to fabri-
cate an atomic bomb similar to those used to destroy Hiroshima and Nagasaki at the end of the Second World War.   In
fact, the desire to have such bombs is undoubtedly a prime
reason that some countries want nuclear energy plants.

To understand the full scope of the hazards of nuclear
plants, both in the present type of plant and in possible
future breeders, it is helpful to realize that when a 1000-
megawatt commercial plant has been running for a while it
contains in its sealed fuel rods a larger quantity of radioactive fission products than is produced in the explosion of
2000 Hiroshima bombs. The American Physical Society's 1974
Reactor Safety Study[13] implies that uniform distribution of
the volatile long-lived Cesium-137 present in a 1000-megawatt
power plant would contaminate 120,000 square miles and ren-
der the area unsuitable for agriculture.[14] Although such
uniform distribution is unlikely, a major release could be
carried downwind for several hundred miles with the heaviest
fallout occuring closer to the plant.

Normally, the fission products from a nuclear reactor
are contained within the sealed reactor fuel rods. The
spent fuel rods are stored under water in a swimming-pool-
like structure next to the building which contains the reactor. Such pools have a capacity to store seven times as
much spent fuel as is in the reactor at any one time. When
and to where this waste will be moved must await a decision
by the federal government of how to dispose of it safely.
One wonders what might result if terrorists blew such a
storage pool sky high.

When the accident occurred at Three Mile Island, the
reactor had not yet reached equilibrium, so there were only
about 300 times as much of the long-lived fission waste
present as was released by the Hiroshima bomb. As a result
of the accident it appears that over 90 percent of the fuel
elements in the reactor were ruptured, releasing some of
the fission products into the cooling water and subsequently
into the containment building. The coming four-year cleanup of this lethal mess is by far the most hazardous part
of the accident. Tentative plans call for hauling the resulting waste, 2000 truckloads of it, to Hanford, Washington,
a distance of 2,500 miles.[15]

What these examples show is that the full impact of
nuclear power is wide-ranging. Potential releases of radioactivity extend over the whole fuel cycle, from mining to
reactor to waste disposal. Exposure of humans, both in gen-

eral and in an occupational setting, can occur at any stage,
not only as a result of accidents, but also as a result of
"routine" operation. It is this cumulative impact of which
we must be aware if we seek to protect our children and
grandchildren from harm.

The Special Lessons of Three Mile Island

Many of the aspects of nuclear power I have reviewed
here are not new, but are discussed at length in the literature of this most difficult subject. Yet beyond these
things, I believe I learned from my service on the Kemeny
Commission some lessons which seem unique and special and
hence worth recounting here.

One of these special lessons concerns an old problem
that plagued scientists at the time of the Inquisition, and
that led to suppression of Galileo and Lavoisier, as well as
countless lesser scientists since then. Why is it that a
Ph.D. nuclear physicist who has worked on nuclear energy in
a highly creditable way with his colleagues for many years
loses his credibility at the moment that he questions the
safety of nuclear energy? I tried to arrange for just one
such nuclear physicist to work side by side with, or just
consult with the nuclear scientists on the Kemeny Commission
staff, who as a group had the mindset that nuclear energy
is safe. This effort of mine was successfully resisted.
Certainly you and I as members of a society dedicated to
the advancement of science should go out of our way to ex-
pose our convictions and prejudices to the views of our
critics.

Another special lesson I learned from the TMI experience is not to be lulled into complacency by the promotional
efforts of the nuclear industry. "Nuclear energy is safe--
period" has been the theme song of the industry for twenty
years. I shall never forget the day we commissioners visited
the undamaged TMI-1 plant. Dressed in two layers of protective clothing with our cuffs and sleeves taped to keep out
contaminants, with two radiation dosimeters on us to measure our exposure, with signs warning us to stay out of contaminated areas, we assembled around a desk while the guide
explained how the massive containment building is designed
to contain the deadly fission products that might escape
from the reactor vessel. On the desk was a sign: "Nuclear power is safe."

Another special lesson which I learned just recently is
that the nuclear industry has not learned its lesson. In
a recent advertisement in major newspapers, the Edison

Electric Institute proclaimed: "The electric companies agree with the Kemeny Commission's message on nuclear power: proceed, but proceed with caution." The Kemeny Commission said no such thing! The introduction to our report carefully explains that our assignment limited us to matters bearing directly on Three Mile Island, and that we made no judgments on a number of other issues, e.g., the disposal of radioactive wastes, that must be considered in deciding whether nuclear power is basically safe. Therefore, our report says: "We did not attempt to reach a conclusion as to whether, as a matter of public policy, the development of commercial nuclear power should be continued or should not be continued." The Commission during its deliberations considered several resolutions calling for a moratorium on construction permits for new nuclear power plants. Although none of them won the seven votes needed for approval, eight of the twelve of us voted for such a moratorium on at least one of these resolutions.

Conclusion

In the preceding discussion I have offered you a number of lessons drawn from the experience at TMI and the operation of nuclear power generally. These indicate that we, as a society, must be more mindful of human error in the operations of nuclear power plants; that we must pay more attention to the widespread fear of nuclear power; that we must take into account the cumulative effects of the nuclear fuel cycle; and that we ought to be careful to give the critics their due, not only at the barricades and in court, but on Presidential commissions, so that we are not lulled into complacency by the dogmatic optimism of the nuclear industry.

Beyond these issues I believe there is one risk worth mentioning, and it is the risk of not having an alternative to nuclear energy, or more specifically, not having one in time. The alternatives are currently quite clearly defined. What is needed now are the resources to develop them. First of all are conservation and the more efficient use of energy. They will permit us to get by with a healthy economy and still use no more total energy in the year 2000 that we are using today. Coal and the family of resources energized by the sun appear as two good prospects to provide us with alternative choices for energy supply. The key to all this is technology. It brings with it benefits and risks. We must uncover the choices available and then assess their long-term potential impacts on all aspects of life, and make our selection so as to maximize the positives and minimize the negatives.

References and Notes

1. U.S., President's Commission on the Accident at Three
    Mile Island, *The Need for Change: The Legacy of TMI*,
    (Government Printing Office, Washington, October, 1979).

2. U.S. Nuclear Regulatory Commission, Special Inquiry
    Group, *Three Mile Island: A Report to the Commission-
    ers and to the Public* (Nuclear Regulatory Commission,
    Washington, January, 1980).

3. U.S. Atomic Energy Commission internal files on 1964-65
    secret update of the 1957 AEC report (WASH 740). This
    updated report was never finalized, but was released
    to the public as part of a freedom of information suit
    in 1973. The report confirmed the predictions made in
    WASH-740, of large possible consequences from reactor
    accidents.

4. U.S. Nuclear Regulatory Commission, *The Reactor Safety
    Study*   (WASH-1400, NUREG 75/014; Nuclear Regulatory
    Commission, Washington, 1975).

5. Louis Harris and Associates, 4-5 April 1979. See also
    "TMI Aftermath: Polling the Public," *Nuclear News*,
    22 No. 7 (May, 1979), 32; R.C. Mitchell, *The Public
    Response to a Highly Publicized Major Failure of a
    Controversial Technology* (Discussion Paper D-60;
    Resources for the Future, Washington, 20 February
    1980).

6. M.J. Schneider, *Persistent Poisons: Chemical Pollu-
    tants in the Environment* (New York Academy of Sciences,
    New York, 1979).

7. R.A. Jones, "Rain Forests Shriveling as Man Intrudes,"
    Los Angeles Times, 7 May 1979.

8. A. LaBastille, "Heaven, Not Hell," *Audubon Magazine*, 81
    no. 6 (November 1979), 68-103.

9. U.S., Interagency Review Group on Nuclear Waste Manage-
    ment, *Report to the President*... (TID-29442; Washington,
    March 1979), pp. 80-84.

10. "Radiation: Our Chemical Chickens Have Come Home to
    Roost," Special Report in *Chicago Tribune*, 1-5 April
    1979.

11. M. Strutin, "In His Sky-Blue Waters--Radioactivity,"
     Christian Science Monitor, 8 January 1980.

12. B. Franklin, "Nuclear Plants are Hiring Stand-ins to
     Spare Aides the Radiation Risk," *New York Times*, 16
     July 1979.

13. *Reviews of Modern Physics*, 47, Supplement (1975).

14. Dividing the 6 million Curie inventory of Cesium-137
     (Table XXXIV of reference 13) by the 20 microcurie/
     square meter threshold for Cesium-137 agricultural
     contamination (Table XXXIX, reference 13) results in
     a figure of 300,000 square kilometers (120,000 square
     miles). A similar calculation for Strontium-90 pre-
     sent in a 1000 megawatt core would lead to a contamina-
     ted area figure ten times higher. However, Strontium-
     90 is less volatile than Cesium-137, and according to
     reference 13, one expects about ten times strontium to
     be released in a catastrophic meltdown. Consequently,
     the expected agricultural contamination from strontium
     and cesium would be comparable.

15. Bechtel Power Corporation, *Recovery Plan for General
     Public Utilities* (Bechtel, San Francisco, 13 August
     1979).

5. Three Mile Island: Lessons for Government

At a few seconds after 4 a.m. on March 28, 1979, nuclear power reached a crossroads. One fork of that road leads to the elimination of nuclear power as a viable energy source in the United States; the other may result in its retention as an energy option. Which road the industry takes is largely dependent on two factors: First, can the industry realistically and substantially change, i.e., will the fundamental changes called for by the President's Commission on the Accident at Three Mile Island (TMI) be carried out effectively? Or will the industry merely make cosmetic changes as part of a public relations campaign to enhance its public image? Second, can government adequately regulate this potentially dangerous energy source so that in the future the public health and safety will be protected?[1] The first issue has been addressed in a companion piece by Russell Peterson, who suggests that the approach of industry in the post-TMI period is little changed (Chapter 4). Here, I shall concentrate on the second issue.

Government must establish itself as a responsible and effective regulator. Industry cannot be relied upon as the principal guarantor of safety. One need only look at the Commission's findings on the role of the utility and its suppliers to demonstrate that fact. In order to improve the government's nuclear regulation role, fundamental changes are necessary. The maze of paper imperatives, guides, rules and regulations that earmark the world of the Nuclear Regulatory Commission (NRC) must reflect the reality of safety; objectives must be well-defined and enforcement must be a primary and visible part of the regulatory scheme. Government regulation must be properly

---

The views expressed herein are solely those of the author. They do not necessarily represent the views of the President's Commission on the Accident at Three Mile Island or any member thereof, or the United States Department of Justice.

managed and must be unequivocally committed to safety. The
investigations of the President's Commission suggested that
the NRC was promotion-oriented. Promotion of nuclear power
should play no role in the regulatory process. (In the following I look at each of these propositions in more detail.)

The Reality of Safety: Siting
and Emergency Planning

The absence of realistic precautions in nuclear safety
regulation can be illustrated by looking at the siting of
Three Mile Island and the emergency plans that existed at
the time of the accident. The NRC required Metropolitan
Edison, the utility that managed TMI, to have an emergency
plan for protective actions only within a two-mile radius
of the plant, the so-called low population zone[2] even though
the plant was only ten miles from Harrisburg, Pennsylvania,
a significant population center. The State of Pennsylvania
required plans for only a five-mile radius around a nuclear
plant. During the accident, local and federal officials
hurriedly tried to develop ten- and twenty-mile evacuation
plans destroying any credibility that previously attached
to the low population zone concept and siting criteria.[3]
Moreover, potassium iodide, an approved agent for protecting
the thyroid gland from radioactive iodine, was not commer-
cially available in the United States in quantities sufficient for the population in the TMI area.[4] In short, before
TMI, government authorities had not realistically assessed the
consequences of locating Three Mile Island near a major population center and had not thought through what was really necessary to protect that population in the event of an accident.

These problems of siting and emergency planning are not
unique to TMI. In testimony before the President's Commission,
Robert Ryan, Director of the NRC's Office of State Programs,
testified about New York's Indian Point nuclear complex:

> I think it is insane to have a three-unit reactor
> on the Hudson River in Westchester County, 40 miles
> from Times Square, 20 miles from the Bronx. And
> if you describe that 50-mile circle...you've got
> 21 million people, and that's crazy.... I just
> didn't think that that's the right place to put
> a nuclear facility.... It's a nightmare from the
> point of view of emergency preparedness.[5]

Clearly, it would have been wiser if the NRC, or the
Atomic Energy Commission (AEC) had thought about these problems before licensing those plants. But that meant looking
at the real world, rather than the fairy-tale exercises

that led plants to be sited at those locations and to emergency plans that had to be thrown out the window when a real accident occurred.

It is the conclusion of the Commission that siting and emergency planning policies must change. Specifically, the Commission has recommended that new plants be sited, to the maximum extent feasible, in areas remote from population concentrations;[6] that potassium iodide should be available for distribution to the general population and workers in a radiological emergency;[7] that emergency plans should be based on technical assessments of various classes of accidents that can take place at a given plant, rather than on a fixed set of distances and a fixed set of responses.[8] Finally, plans must be workable. The Commission recommended that a utility should not be granted an operating license until the state in which the plant is to be located has an emergency response plan reviewed and approved by the new Federal Emergency Management Agency.[9]

The Reality of Safety: NRC Licensing Reviews

A lack of reality also permeates the licensing process for nuclear power plants. The President's Commission found serious inadequacies in the NRC's licensing process. To illustrate, I will consider three examples drawn from the work of the President's Commission.

(a) *License applicants need only analyze "single failure" accidents and are not required to analyze what happens when systems fail independently of each other.*[10] Yet the TMI accident was a multiple-failure accident--both the main and auxiliary feedwater systems failed and the pilot-operated relief valve (PORV) stuck open.[11] Indeed, TMI, according to the NRC staff, was a class 9 accident,[12] a sequence of failures more severe than those considered in the design basis for protective systems and engineered safety features. Such a sequence is thought to be of extremely low probability.[13] Realistic multiple-failure analysis--including human failures--clearly must be an ingredient of future licensing reviews if a greater degree of safety is to be assured.

(b) *The NRC generally restricts its design review to those items considered "safety-related." Non-"safety-related" components are not subjected to the same level of analysis required for components designated "safety-related."*[14] That arbitrary and artificial distinction--a distinction apparently based on historical rather than logical reasons[15]--

directly impacted the TMI-2 accident. The pilot-operated relief valve that stuck open during the accident was not considered "safety-related" because it had a block valve to
isolate it from the primary system, and the block valve was
not considered "safety-related" because it had a PORV in
series with it.[16] This rationale is all the more incredible
when one considers that between 1971 and March 28, 1979,
PORVs had stuck open at least 11 times in pressurized water
reactors, and 9 of those 11 failures had occurred in Babcock
and Wilcox reactors.[17] In fact, one of those failures had
occurred at TMI-2 in 1978, a year before this accident, when
the valve failed to open upon a loss of power. Even though
an NRC inspector requested a review of the design approach,
the NRC blithely took the position that no major action was
necessary because the TMI-2 high pressure injection--the
fundamental safety system that the operators effectively
turned off on March 28, 1979--would be more than able to
compensate for any water lost through a stuck-open PORV.[18]

(c) *The NRC licensing review excludes consideration of
the person-machine interface.* Thus, control-room instruments are arranged in an awkward fashion, especially for
quick response to an accident; operators are merely taught
the test in order to qualify for an NRC license; and plant
operating and emergency operating procedures are sometimes
devised and implemented even though they may be hard to
follow, or, in the case of certain TMI-2 emergency procedures, incomprehensible.[19] The potential for disaster
inherent in each of these deficiencies in the event of an
accident apparently did not enter into the process, because,
I honestly believe, no one thought an accident would ever
really occur.

(d) *The Advisory Committee on Reactor Safeguards (ACRS),
the statutory expert committee with authority to advise the
NRC on safety aspects of proposed and existing plants, has
a very limited role.* There is some support for the proposition that ACRS concerns are ignored by the NRC. During the
TMI-2 operating license review, ACRS questioned Metropolitan
Edison's failure to delineate safety-related responsibilities for TMI-2 personnel.[20] In response, the utility merely
submitted a statement setting out the composition of its
plant staff. Although this response did not address the ACRS
concern about safety-related responsibilities, the NRC staff
considered the matter "resolved."[21] The NRC staff undoubtedly came to this conclusion either because they believed
that since nuclear plants are safe, accidents really cannot
happen so there is no need for the safety function the plant
staff was supposed to undertake--an assumption that defies
reality--or they never bothered to read the response.

In 1977, during the licensing review for the Pebble Springs plant, ACRS asked the applicant about operator ability to interpret correctly the pressurizer level after a relatively small pipe break leading to a loss-of-coolant accident. Although the Reactor Systems Branch of NRC was assigned responsibility for assuring that this question was answered, it was not, and that failure to answer was not questioned.[22] During the TMI-2 accident, misleading pressurizer level indication at the very least confused the operators and led to incorrect operator actions.[23] Again, an ACRS concern was ignored.

These and other deficiencies in the NRC licensing process betrayed a very limited view of nuclear safety needs. Failure analysis demanded of industry was one-dimensional and unrealistic, the interaction of people with instruments was ignored, and the advice of experts on reactor safety was treated in a perfunctory manner, if at all. These deficiencies directly impacted the accident at TMI-2. Underlying the approach can only be the belief that nuclear plants are safe, that accidents cannot really happen, and that the most important goal of regulation and licensing is to facilitate development of the technology. These deficiencies must be corrected if the nuclear licensing process is to be oriented realistically toward safety.

Lack of Effective Regulation: Inspection and Enforcement

Effective regulation requires an effective inspection and enforcement capability. It makes little difference what rules and regulations are promulgated unless the industry firmly believes that a failure to comply will result in timely and meaningful action by the regulator. The evidence generated by the President's Commission clearly indicates that NRC's Office of Inspection and Enforcement (I&E) does not fit that description. Instead, it was at the time of the TMI accident lax, uneven, and therefore ineffective. Several examples amplify this point.

(a) *In 1978, the General Accounting Office (GAO) reported that NRC inspectors rely unduly on utilities and their suppliers to determine whether plants are being adequately constructed.*[24] That undue reliance carries over into the monitoring of operating plants. In 1977, an incident occurred at Toledo Edison's Davis Besse I. At Davis Besse, apparently relying on a misleading pressurizer level indication, the operators terminated the high pressure injection safety system after a PORV stuck open. Yet, Toledo's reports of the event either failed to mention this fact or concluded that "operator action was timely and proper throughout the sequence of events."[25]   I&E's final inspection report on the incident echoed this almost nonsensical aura of normality by merely stating: "HPI pumps were shut down...as pressurizer level was normal."[26]   Another example is Metropolitan Edison's failure to report the procedure, begun in August 1978, of closing the now famous auxiliary feedwater block valves at TMI-2 even though the closing violated the TMI-2 Technical Specifications.[27] NRC did not discover that violation until after the TMI accident.

(b) *NRC's Inspection and Enforcement office has failed to evaluate systematically the operating data it receives from plants.*   This is a most serious failure, since criteria for inspection and enforcement cannot be meaningful in the absence of such information.   Simply put, the Licensee Event Report system is a mess.   NRC recognized this problem, but nothing was done before the accident at TMI-2,[28] apparently in part because of a jurisdictional dispute between I&E and the Office of Nuclear Reactor Regulation.[29]   Indeed, there is some question in my mind as to whether NRC has yet learned this lesson from TMI.   On September 25, 1979, an incident occurred at North Anna I (Virginia) that led to activation of that plant's high pressure injection system. During the first minutes of the incident, the North Anna control room was in contact with an NRC employee in the NRC's Incident Response Center who was making a routine status check on the direct-line telephone system installed in each plant as a result of the TMI accident.   When the NRC employee was informed that high pressure injection, the essential safety system involved in TMI-2, had been activated, he said "Fine, thank you," hung up the telephone, and went about his business.[30]   That incident raises the fundamental question of whether adequate information-gathering, evaluation, and dissemination is taking place in NRC even after TMI.

(c) *I&E enforcement activities have left much to be desired.   In a February 1979 report, GAO found that NRC had not made full and effective use of its present enforcement authority.*[31]   GAO cited case studies for its conclusion, including the fact that NRC headquarters downgraded civil penalties proposed by I&E regional offices.   In particular, NRC limited its response to enforcement letters and meted out penalties much smaller than those called for by its own enforcement procedures because of "perceived but unsubstantiated licensee financial hardship."[32]

As in the case of licensing, NRC inspection and

enforcement do not measure up. The regulator's inspection
and enforcement arm must act firmly and effectively to be
credible. A meaningful enforcement effort coupled with a
systematic inspection system can serve to deter violations
that might otherwise take place. To improve this situation
the President's Commission has called for marked improvement in this area, including a periodic intensive and open
review of each licensee's performance according to the requirements of its license and applicable regulations.
Licensees found to be substandard during intensive review
would be subject to sanctions, including license revoca-
tion.[33]

NRC Management

Let me turn for a moment to the issue of the NRC's attitude toward regulation. NRC is still "fundamentally geared
to trying to nurture a growing industry."[34] This reflects
a "mindset" that was theoretically eliminated by the Energy
Reorganization Act of 1974. The NRC Commissioners have
largely isolated themselves from the licensing process
through the use of unnecessarily broad rules that prohibit
the Commissioners from discussing pending matters with the
staff.[35] Moreover, they have isolated themselves from the
overall management of the agency. According to one present
NRC Commissioner, "There has been too little Commission
involvement in the setting of policy in this agency and
little Commission guidance on safety matters to the staff."[36]
This promotional philosophy, coupled with the fact that the
agency is unmanaged rather than mismanaged, creates a regulatory paralysis that often favors the licensee in disputes
involving public health and safety. Let me offer two examples:

(a) *The NRC has been reluctant to order "backfitting,"
that is, applying a new requirement retroactively to a plant
that has already received its license.*[37] In May 1977, NRC
adopted a new Regulatory Guide, upgrading requirements for
licensee emergency plans.[38] At the time of the TMI accident, however, those requirements had been met by only four
plants--all licensed after the Regulatory Guide was adopted.
NRC's failure to backfit these emergency plan requirements
makes little sense since it imposes a minimal financial burden on the industry. In light of TMI, NRC is apparently
reevaluating this indefensible policy.[39]

(b) *Owing to its loose organization the NRC's actions
during the accident were contradictory and ineffective and
often resembled those of the Keystone Cops rather than those
of a responsible regulatory body.* The NRC staff recommended

evacuation to the Governor of Pennsylvania on Friday,
March 30, based on insufficient information and without
even consulting the Commissioners.[40] In fact, NRC's emergency plans did not contemplate specific Commissioner involvement in NRC emergency actions.[41] Moreover, the Commission acted as a collegial body during the emergency--no
one person was clearly in charge.[42] Finally, the hydrogen
bubble debacle does little to inspire confidence in the
agency's technical ability to proceed rationally and systematically in a crisis situation.[43]

    To improve on both routine and crisis management, the
President's Commission has recommended eliminating the Commission structure and replacing it with a single administrator. The single administrator would have substantial dis-
cretionary authority over the internal organization and
management of the agency. Coupled with and as an integral
part of that restructuring, the President's Commission
recommended the creation of an outside oversight committee
to examine the performance of the NRC and the industry on
an ongoing basis, and to report its findings annually to
the President and the Congress.

    This proposal has generated some debate. But we must
candidly recognize that the 1974 creation of the NRC was
nothing more than a shuffling of boxes. As Coates (Chapter
3) has noted, the regulatory area of the AEC was simply
renamed and thus continued along the same course it had followed before the shuffle. The President's Commission has
now suggested radical surgery, if necessary, including the
elimination of senior staff personnel who can not or will
not realize that their allegiance must be to safety regulation rather than "nurturing a growing industry." In
response, President Carter had recommended that the powers
of the NRC Chairman be strengthened and his role clarified.
This clearly represents a compromise between the old system
and the recommendations of the President's Commission. One
hopes that it will be an effective compromise, and that
the new, improved role of the chairman will enable him to
take hold of the NRC and redirect its effort toward a more
consequential approach to safety.

Summary and Conclusion

    In the preceding discussion I have shown by examples
drawn from the work of the President's Commission on the
Accident at Three Mile Island that the NRC has done an inadequate job in regulating the nuclear industry and providing
the public with safe nuclear power. This inadequacy of the
NRC is rooted in a predisposition to believe that nuclear

power is safe. It is expressed through a lack of realism in siting and emergency planning; through failure to take licensing reviews sufficiently seriously; through lack of adequate inspection and enforcement. Beyond this, the NRC suffers from a diffuse organizational structure that is independent of specific regulatory shortcomings. This structure promotes irresponsible decision-making and buck-passing, with the eventual result that public safety is impacted.

In response to this situation the President's Commission concluded that

> with its present organization, staff, and attitudes, the NRC is unable to fulfill its responsibility for providing an acceptable level of safety for nuclear power plants.44

Based on this finding the President's Commission has called for fundamental changes. Chief among these is reorganization of the NRC, and removal if necessary of senior staff who hold a viewpoint that goes back to the old "promotional" attitude of the Atomic Energy Commission. But reorganization is only a beginning. If the NRC is to work, what must ultimately happen is the elimination of many of the deficiencies that produced the present inadequate record of the NRC.

Although critics of the NRC tend to assume that whatever needs to be done can be done, it may not be so easy. Nuclear power is among the most unforgiving of technologies, and in the view of Weinberg requires more intensive watchfulness by society than any comparable enterprise.45 The continuing failure of the NRC, if it persists, may be rooted as much in the exacting requirements of the technology as in inadequate regulatory structure and procedure.

Whatever the future may hold, Three Mile Island was a turning point that brought public confidence in the NRC to a new low. It demands a new departure, a new blueprint for safety regulation, which if not successful, may contribute to the demise of nuclear power as a viable energy option. What is needed for the short run is rapid and effective institution of change, along the lines suggested by the President's Commission. This should create an atmosphere of competent and consequential regulation. Failing in this will mean that government has learned little from the events at Three Mile Island. Moreover, once committed to effective regulation, government must maintain that commitment by erecting and retaining the thickest possible wall between promotion of the technology and safety regulation.

References and Notes

1. The President's Commission found that the major health effect of the TMI accident appears to have been on the mental health of the people living in the region of Three Mile Island and of the workers at TMI. See U.S. President's Commission on the Accident at Three Mile Island, *The Need for Change: The Legacy of TMI*, *Report of...*   (Government Printing Office, Washington, 1979), p. 35 (hereafter cited as *Report*).

2. The low population zone or LPZ is defined as an area containing "residents, the total number and density of which are such that there is a reasonable probability that appropriate protective measures could be taken in their behalf in the event of a serious accident," 10 *CFR* §100.3(b) (1979). For a discussion of LPZ's, see U.S. President's Commission on the Accident at Three Mile Island, *Report of the Office of the Chief Counsel on Emergency Preparedness* (Government Printing Office, Washington, October 1979), pp. 3-6 (hereafter cited as *Emergency Preparedness*).

3. Indeed, the adequacy of emergency planning for TMI-2 had been attacked during the operating license stage for TMI-2. Contention Number 8 raised by the intervenors to that proceeding stated:

    > The warning and evacuation plans of the Applicants and the Commonwealth of Pennsylvania are inadequate and unworkable. The plans assume that all local and state officials involved are on 24-hour notice and can be contacted immediately. They further assume that all people notified will promptly react and know how to respond and are trained in what to do. They also assume that the public, which has been assured that accidents are "highly unlikely" or "highly unprobable," will respond and allow themselves to be evacuated. No operating license should be granted for Unit 2 until emergency and evacuation plans are shown to be workable through live tests.

    See *Emergency Preparedness*, pp. 11-12. The population within a 10-mile radius of TMI in 1970 was approximately 121,000; by the year 2000 it is estimated to be 183,000. See U.S. Nuclear Regulatory Commission, *Beyond Defense-in-Depth* (NUREG-0553; NRC, Washington,

October, 1979), p. II-90 (hereafter cited as *Beyond Defense-in-Depth*).

4. *Report*, pp. 41-42; U.S., President's Commission on the Accident at Three Mile Island, *Technical Staff Analysis Report on Public Health and Epidemiology* (Government Printing Office, Washington, 1979), p. 73.

5. Ryan deposition, p. 71. Mr. Ryan believes that if the TMI accident had occurred at Indian Point loss of life might have resulted had an evacuation occurred. *Ibid.*, pp. 71-72.

6. *Report*, p. 64. According to the NRC, the Report of its Siting Policy Task Force, being considered by the NRC Commissioners at the time these remarks are being written, "recommends changes which parallel, and in some cases go beyond, those of the President's Commission." Letter from Joseph M. Hendrie, Chairman, NRC, to Dr. Frank Press, Director, OSTP (9 November 1979), Enclosure 1, p. 5 (hereafter cited as "Hendrie Letter").

7. *Report*, p. 75. NRC agrees with the recommendation as well. "Hendrie Letter," p. E-2.

8. *Report*, p. 76. In October 1979, NRC approved the concept of Emergency Planning Zones of 10 miles for protective actions aimed primarily at reducing whole-body exposure from the plume and from deposited material and 50 miles for protective actions relating to ingestion exposure pathways. See *Beyond Defense-in-Depth*, p. II-84; The NRC statement does not specify that these recommendations will be applied to existing plants. U.S. Nuclear Regulatory Commission, NRC Policy Statement: "Planning Basis for Emergency Response to Nuclear Power Reactor Accidents" (undated, although issued 5 October 1979).

9. *Report*, p. 76.

10. For a definition of "single failure" see 10 *CFR* Part 50, Appendix A (1979).

11. Ross deposition, p. 119; U.S., President's Commission on the Accident at Three Mile Island, *Report of the Office of Chief Counsel on the Nuclear Regulatory Commission* (Washington, October, 1979), pp. 105-108 (hereafter cited as *NRC Report*).

12. Matter of Public Service Electric & Gas Co. (Salem Nuclear Generating Station, Unit 1). Docket 50-27 (24 August 1979).

13. Proposed annex to Appendix D, 10 *CFR* Part 50, *Federal Register,* 36 (1 December 1971), 22852.

14. *Report,* p. 52; *NRC Report,* pp. 95-99; U.S. Nuclear Regulatory Commission, *TMI-2 Lessons Learned Task Force Final Report* (NUREG-0585; NRC, Washington, October, 1979), p. 3-2; U.S. Nuclear Regulatory Commission, *TMI-2 Lessons Learned Task Force Status Report and Short Term Recommendations* (NUREG-0578; NRC, Washington, July, 1979), p. 17.

15. Mattson deposition, p. 97.

16. *Ibid.,* p. 98, TMI Commission hearings, Mattson Testimony, (22 August 1979), p. 240.

17. U.S., President's Commission on the Accident at Three Mile Island, *Technical Staff Analysis Report on PORV Design and Performance* (Government Printing Office, Washington, 1979).

18. Seyfrit deposition, 62-63; *NRC Report,* pp. 106-107.

19. *NRC Report,* pp. 108-121.

20. Letter from Dade W. Mueller to Marcus A. Rowden (22 October 1976), p. 3-4.

21. Safety Evaluation Report, TMI-2 Supp. 2. Feb. 1978, pp. 18-3.

22. *NRC Report,* pp. 76-77.

23. See *Report,* p. 22; U.S., President's Commission on the Accident at Three Mile Island, *Report of the Office of Chief Counsel on the Managing Utility and its Suppliers,* (Government Printing Office, Washington, October, 1979), pp. 102-123.

24. U.S. General Accounting Office, *The Nuclear Regulatory Commission Needs to Aggressively Monitor and Independently Evaluate Nuclear Power Plant Construction* (EMD 78-80; GAO, Washington, 7 September 1978), pp. iii,5.

25. Toledo Edison Company, *Supplemental Report,* 14 November 1977, pp. 2,4. See generally *NRC Report,* pp. 156-160.

26. *Inspection Report* No. 50-346/77-32, 21 November 1977.

27. TMI-2 Technical Specification 3.7.1.12; *NRC Report*, p. 160.

28. See *NRC Report*, pp. 175-181.

29. Budnitz deposition, pp. 62-63.

30. *NRC Report*, p. 208, note 696.

31. U.S. General Accounting Office, *Higher Penalties Could Deter Violation of Nuclear Regulation* (EMD-79-9; GAO, Washington, 16 February 1979), pp. 10,20.

32. *Ibid.*, pp. 12, 18.

33. *Report*, p. 66.

34. Ahearne deposition, p. 230; *Report*, p. 51.

35. *NRC Report*, pp. 34-44.

36. Victor Gilinsky, testimony in U.S. Congress, House, Committee on Interior and Insular Affairs, Subcommittee on Energy and the Environment, *Nuclear Regulation: Hearings...June 4 1979* (96th Cong., 1st sess.; Washington, 1979), pp. 74-75; also V. Gilinsky, testimony in U.S. President's Commission on the Accident at Three Mile Island, *Hearings...June 1, 1979;* pp. 143-144.

37. 10 *CFR* §50.109(a)(1979); see *NRC Report*, pp. 123-134.

38. U.S. Nuclear Regulatory Commission, Office of Standards and Development, *Regulatory Guide*, 1.101 (May, 1977).

39. Memorandum from Harold Denton to NRC Commissioners (SECY-79-450, 23 July 1979), Enclosure 1, p. 1.

40. U.S., President's Commission on the Accident at Three Mile Island, *Report of the Office of Chief Counsel on Emergency Response* (Government Printing Office, Washington, 1979), pp. 41-50 (hereafter cited as *Emergency Response*).

41. Hendrie deposition, p. 231.

42. *NRC Report*, pp. 45-48, 27-33.

43. See *Emergency Response,* pp. 110 ff.

44. *Report,* p. 56.

45. A. Weinberg, "The Maturity and Future of Nuclear
     Energy," *American Scientist,* 64 no. 1 (January/Febru-
     ary, 1976), 21.

6. Love Canal: A Review of Government Actions

The dream of growth, prosperity, and progress in the 1800s had become by the 1970s an exploding environmental time bomb. It all started in 1892, when William T. Love, a man of remarkable talent and vision, set out to develop a model industrial city of 600,000 on 47 square miles of land, to be supported by a seven-mile navigation/power canal connecting the upper and lower levels of the Niagara River near Niagara Falls. Love nearly succeeded. He obtained from prominent people and legislative authorities support to comdemn property and divert an unlimited quantity of water from the Niagara River. He actually dug 3,000 feet of his planned canal, but the economic depression of the 1890s and the discovery of alternating current (making possible long-distance transmission of power) brought the project to an end. The partially completed canal, 80 feet wide and 30 feet deep, became for a time a popular swimming hole for children.[1]

The trouble began in the 1920s, when the canal became a waste disposal site for several chemical companies and the City of Niagara Falls. Such practices were continued through 1953, when Hooker Chemical, then owner of the site, deeded it to the Niagara Falls Board of Education for $1, with a warning that potentially hazardous wastes were present. Chemical wastes disposed of at the site are not exactly defined by available records, but are reported to include chemical process residuals and sludges, fly ash, and other industrial waste.[2] In the late 1950s, despite the city's knowledge of the buried wastes, development of the site began directly adjacent to the landfill. About 230 homes and an elementary school were built along two streets (97th and 99th) which ran parallel to the canal on either side (See map, Figure 6.1).

All went well for a while. Recently, however, ponded surface water, heavily contaminated with chemicals, emerged

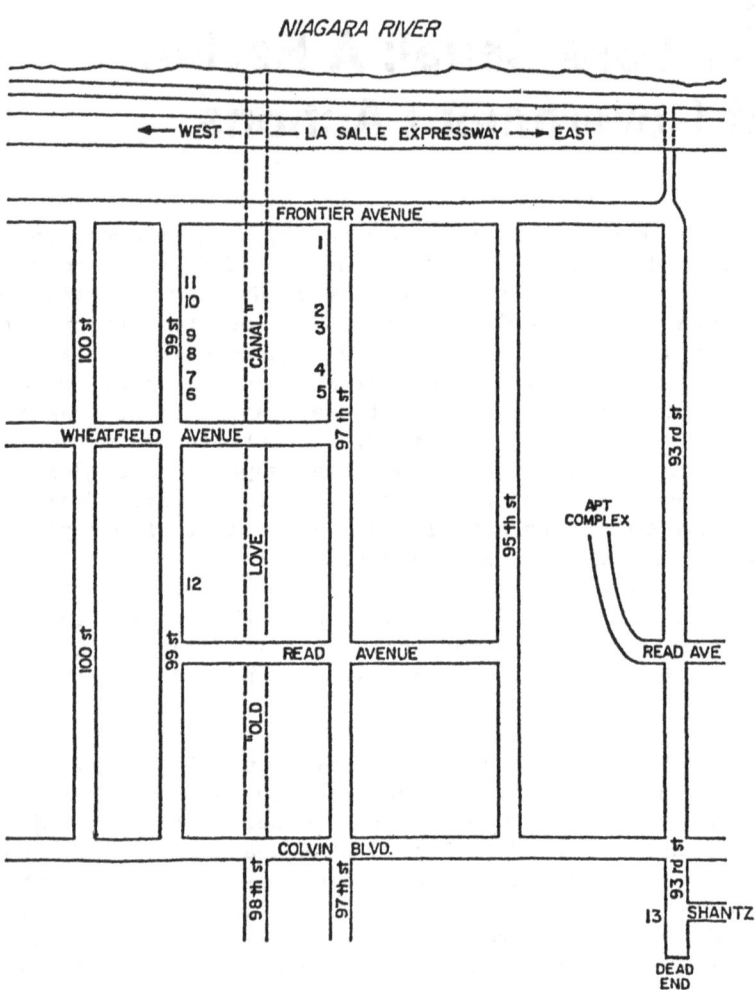

Figure 6.1. Map depicting sampling locations 1 to 13 in Niagara Falls, New York.

in several back yards. Unpleasant odors outdoors and in
basements became prevalent, especially in warm weather.
Local residents found themselves replacing sump pumps because of the corrosive nature of liquids being collected
in these pumps. Noxious fumes and potentially hazardous
liquids were detected in several storm sewers. School personnel reported to the county health department that children playing in the school yard had received skin burns.
In 1977, prolonged rainfall and one of the worst blizzards
ever to hit the area probably caused significant increase
in migration of contaminated leachate, producing a significant rise in the level of local complaints.[3] These and
government concern finally triggered a combined federal,
state, and local effort to define the extent of the problem,
to assess the health and environmental implications, and to
develop remedial measures. In the remainder of this paper
I provide a review of the government's action, particularly
the steps taken by the Environmental Protection Agency
(EPA).

Government Investigation of the
Degree of Contamination

Early in 1976, when the Love Canal situation was first
brought to the attention of the Federal Government, it was
determined that the New York State Department of Environmental Conservation (NYSDEC) had sufficient legislative
authority to handle the situation. At the time, no federal
legislation regarding hazardous waste landfills was in existence. The Resource Conservation and Recovery Act (RCRA)
was not passed until later that year.

In April 1977, the Calspan Corporation of Buffalo, under
contract to the City of Niagara Falls, initiated studies
to assess the extent of chemical contamination and to propose methods for control of leachate migration.[4] These
studies, which in the spring and summer of 1977 involved
sampling of basements and sewers and the construction of
six monitoring wells, suggested the possible presence in
sumps and sewers of several hazardous constituents, including polychlorinated biphenyls, hexachlorobenzene, and hexa-
chlorocyclopentadiene. Several remedial actions were considered, including the construction of a tile drain system
to intercept leachate, wellpoints into the landfill to pump
water from the site, chemical fixation, and excavation and
removal of wastes to an environmentally secure site. The
Calspan study, while leaning toward the tile-drain system
as the most appropriate approach, suggested that additional
sampling would be needed before action could be taken.

Table 6.1. Household basements: organic chemicals
identified in air and water.

Chemical	Air	Sump	Health Significance
Benzene	X	X	Leukemogen
Dichloroethylene	X		Animal Carcinogen
Methylene Chloride	X		Respiratory Distress
Chloroform	X	X	Animal Carcinogen
1,1,1 Trichloroethane	X	X	
Carbon Tetrachloride	X	X	Animal Carcinogen
Trichloroethylene	X	X	Animal Carcinogen
Tetrachloroethylene	X		Animal Carcinogen
Pentachloroethane	X		
Pentachlorobutadiene	X		
1,3 Hexachlorobutadiene	X		Animal Carcinogen
Chlorobenzene	X	X	
Dichlorobenzene Isomers	X		
Trichlorobenzene Isomers	X	X	
Tetrachlorobenzene Isomers	X		
Chlorotoluene Isomers	X	X	
Dichlorotoluene Isomers	X	X	
Trichlorotoluene Isomers	X	X	
Tetrachlorotoluene Isomers	X		
Toluene		X	Narcosis, Possible Anemia Agent
Hexachlorocyclohexane Isomers		X	
Perchloroethylene		X	

Source: Adapted from New York (State), Dept. of Health,
Office of Public Health, Love Canal: Public Health Time
Bomb (Dept. of Health, Albany, 1978), p. 12; New York
(State), Dept. of Health, "Results of Mass Spectrometry
Analysis: Organic Compounds Identified in Three Types"
(Dept. of Health, Albany, n.d.), reprinted in U.S., Congress,
Senate, Committee on Environment and Public Works, Sub-
committee on Environmental Pollution, Hazardous and Toxic
Waste Disposal: Joint Hearings..., pt. 1 (96th Congress,
1st. sess., Serial 96-H9, 1979), p. 216.

In September 1977, further sampling in the area was conducted by the New York State Department of Environmental Conservation. Samples of ponded water analyzed by the New York Department of Health (NYDOH) showed the probable presence of toluene, benzene, dichlorotoluene, trichlorobenzene, and tetrachlorobenzene.[5] Later state monitoring of exposed sludges and ponded water showed the probable presence of lindane analogues, benzene, toluene, chlorobenzene, dichlorobenzene, and trichlorophenol.

In further studies in February 1978, EPA and NYSDEC conducted a basement air and sump water sampling program.[6] Eleven homes in the southern section of the site, and two schools, one on-site and one off-site, were examined over a three-day period, using air samplers, air pumps, and airflow recorders as well as "grab" samples of sump water. The studies indicated the presence of a large number of halogenated organics and benzene. For example, 40 compounds were identified in the air, with estimated levels, while about 40 others were just identified as present. Many of these compounds had serious health significance, as shown in Table 6.1.

After an escalating series of studies over a period of almost two years, it was thus clear that unhealthy conditions existed at Love Canal, and that people in their homes were being exposed to a significant number of toxic chemicals. These included benzene, a known cause of leukemia in humans, and six suspected carcinogens based on animal tests (Table 6.1). In response to this situation, Robert P. Whalen, Commissioner of the New York Department of Health, on April 25, 1978 issued to the County of Niagara Board of Health and the City of Niagara Falls a directive to relieve the immediate hazard.[7]

Subsequently, intensive cooperative studies were initiated by NYSDEC, NYDOH, and EPA. Several compounds were selected for evaluation as indicator parameters. Initial sampling was done in homes directly adjacent to the landfill (ring 1), with particular emphasis on air quality measurement and risk estimation. Soil sampling was carried out to define the limits of the canal and estimate the extent of migration beyond the streets immediately adjacent to the site. Deep-wells were constructed to begin assessment of groundwater contamination. In the later stages of the studies, sampling was extended to homes across the street from ring 1 (ring 2), and finally to homes outside the streets paralleling the canal (ring 3). The hope is to determine the extent and velocity of leachate migration, the

Table 6.2. Concentration of 5 organic chemicals in air samples taken near Love Canal in June-August 1978. (units of micrograms/m^3)

Location	No. Houses	Range of Values	Median	Mean	% with Measurable Level
Ring 1 North 97th	25	0-393	17	67	92%
Ring 1 North 99th	28	0-142	9.5	29	89%
Ring 1 North, Total	53	0-393	.12	47	91%
Ring 1 South 97th	22	0-3616	53.5	427	95%
Ring 1 South 99th	24	0-6944	24	356	96%
Ring 1 South, Total	46	0-6944	28	390	96%
Ring 2 North 97th	22	0-43	0	6	41%
Ring 2 North, 99th	25	0-149	0	12	48%
Ring 2 North, Total	47	0-149	0	9	45%
Ring 2 Central 97th	15	0-69	3	10	67%
Ring 2 Central 99th	13	0-170	0	13	15%
Ring 2 Central, Total	28	0-170	0	12	43%
Ring 2 South 97th	21	0-63	8	13	62%
Ring 2 South 99th	28	0-37	0	4	43%
Ring 2 South, Total	49	0-63	2	8	51%
Ring 1 Total	99	0-6944	17	207	93%
Ring 2 Total	124	0-170	0	9	47%

Source: <u>Time Bomb</u>, p. 9.

size and location of populations at risk, and the effect of remedial measures on water and air levels.

By August 1978 several conclusions could be drawn:

(1) The percentages of contaminated homes in the northern and southern section of ring 1 were similar.

(2) Levels of contamination were higher in ring 1 south than in ring 1 north.

(3) Ring 2 homes were more generally free of contamination than ring 1: i.e. 55% of ring 2 homes and only 5% of ring 1 homes were free of contamination.[8]

Detailed figures are given in Table 6.2.

Epidemiological Studies

Concurrently with the environmental sampling programs initiated by various government groups, the New York Department of Health undertook in June 1978 an epidemiological study to determine the possible increased risk to Love Canal residents. This study was focussed on primary health indicators such as spontaneous abortions, congenital defects, liver function, and blood mercury levels. In addition, blood counts were made to determine possible chronic effects from benzene exposure. The study initially concerned 230 adults and 134 children residing in 97 homes located in ring 1. Personal interviews were conducted to obtain all relevant information on pregnancy history. Where possible and necessary, physician and medical records were searched. State health officials were able to draw a number of conclusions from this work:

(1) The percentage of miscarriages and birth defects was significantly higher than expected for unexposed populations, particularly among women living in the southern portion of ring 1: the data indicated that the risk of spontaneous abortions in Love Canal residents is about 1.5 times greater than that expected for the general population. Detailed examination of data further showed an excess of spontaneous abortions occurring during summer months.

(2) Five congenital malformations occurred among children of adults presently living on Love

Canal. Three were found on 99th Street south,
one on 99th Street North, and one on 97th street.

(3) Women with miscarriages were residents on the Canal
for an average of 18.6 years, while women with no
miscarriages resided on the canal an average of
11.5 years. Age difference or the number of
pregnancies were not statistically significant
factors in the observed differences.[9]

While these findings were not indicative of a conclusive
link between exposure to chemicals and health, they nevertheless provided state and federal experts with significant
information. In 1979 and 1980 this led to further attempts
to link exposure and specific biological effects in the
local population. One such study, conducted under EPA spon-
sorship by Biogenics of Houston,[10] indicated significant
chromosome damage in exposed people, but subsequently came
under attack because of the absence of adequate controls.[11]
Other studies are currently under way at the Center for Disease Control in Atlanta. In addition, a number of privately
sponsored studies have been undertaken by local university
and medical faculty.

Action Is Taken

Despite the preliminary nature of epidemiological studies, evaluation of the available information led the Commissioner of the New York Department of Health, under
authorities contained in Public Health Law, Section 1388,
to declare on August 2, 1978 the existence of a health emergency at Love Canal. In particular he ordered the Niagara
County Health Department and the City of Niagara Falls to
take the following steps.

(1) Remove from the Love Canal site all chemicals,
pesticides, and any other toxic materials on
the surface of the site.

(2) Prohibit and/or severely limit access to the site.

(3) Abate the public nuisance by investigating the
feasibility of corrective actions to lower the
elevated levels of organic compounds in basement air, using moisture proofing and venting
methods.

(4) Undertake the necessary engineering studies
for a long-range solution to decontamination
of the site.

(5) Implement a report by Conestoga-Rovers Associates, entitled "Phase 1: Pollution abatement plan, upper ground water regime," as an interim solution to halting migration of toxic substances (the plan suggested construction of a tile drainage system).

(6) Temporarily delay the opening of the 99th Street School to minimize exposure of school-age children to chemicals while corrective construction takes place.

(7) Evacuate higher risk residents, consisting of pregnant women and children.

(8) Undertake additional studies to delineate chronic diseases infecting all residents who lived in the Love Canal area, with particular emphasis on the frequency of spontaneous abortions, congenital defects, and other pathologies, including cancer; and to determine by appropriate air, water, and ground sampling the extent of leachate migration or whether ground-water aquifers have been contaminated.[12]

With the issuance of this health order, action by government was, in effect, escalated to a critical level. Public concern across the state and throughout the nation began to focus not only on Love Canal, but on other solid waste "time bombs." On August 3, 1978, Governor Carey asked the President to declare Love Canal an emergency.[13] The Governor also established a high-level state task force, chaired by the Commissioner of Transportation, to effect the timely implementation of the health order, particularly the relocation of affected residents, evaluation of the aerial extent of the problem, conduct of health studies, and construction of remedial measures to relieve the emergency. On August 7, 1978 the President issued an emergency declaration to provide Federal resources as needed to protect health and property. Subsequent to this, not only pregnant women and children, but a total of 239 high risk families residing in ring 1 and ring 2 were evacuated from their homes.

Concurrent with the President's emergency declaration, EPA regional administrator Beck was meeting with top level EPA staff and legislators regarding funding for cleanup programs. Beck, as Chairman of the Federal Regional Council, also ordered a review of Federal programs applicable to the problem. This led to the approval for EPA to initiate

in fiscal year 1979 a $4 million state-matching demonstra-
tion grant, designed to begin the clean-up process. In
response, the state legislature appropriated $10.2 million
for purchase of all the evacuated homes in rings 1 and 2,
and $7.8 million for first instance financing of the reme-
dial program anticipated in the EPA grant.

The Remedial Program

A first step of remedial action was the implementation
of the drainage plan proposed to the City of Niagara Falls
by Conestoga-Rovers and Associates.[14] Construction, originally ordered by Health Commissioner Whalen in August,
1978, was completed in January, 1979, using funds made available under the President's emergency declaration. The
action, which involved only the southern portion of the
Canal, consists of a tile drainage collection system to
intercept and transport contaminants for the purpose of
adequate treatment. Included was the placement of an impervious clay cover to eliminate the influence of surface runoff on canal migration. These actions were expected to
lower the water table level, eliminate the outward flow of
contamination, and reverse the flow of ground water away from
surrounding areas back to the Canal. A study by Fred C.
Hart Associates for the EPA confirmed the tile drain system
with clay cover as the principal means of decontamination.[15]

The drain system itself consisted of an eight-inch
diameter perforated pipe, placed 15 feet below the surface,
and located in adjacent back yards to avoid trench work
directly in the highly contaminated landfill itself. The
drains were sloped to wet wells from which the leachate is
pumped into a 30,000 gallon holding tank, then to a special
treatment system on site. After treatment, the effluent
is discharged into the City's sanitary sewage system.

Throughout construction, a comprehensive safety plan
was in effect to protect workmen, residents, and the public
from possible chemical exposure. Developed and carried
out by the state Department of Health, the plan established
site security, work safety procedures, on-site monitoring
to measure in-trench ambient levels, and an emergency evacuation plan to protect the remaining resident population.
In addition, state and city representatives maintained a
close liaison with local residents. Frequent meetings and
issuance of bulletins were commonplace throughout construction.[14,16]

Construction activities produced contaminated liquids
that required proper handling and disposal. Available data,

particularly a laboratory study conducted by Calgon for the
city, suggested that activated carbon might provide effective treatment for leachate collected during construction of
the tile drain system.[17] Plans were thus made to apply this
technique at the site. The EPA Mobile Physical/Chemical
Treatment System, "Blue Magoo,"[18] was initially employed to
demonstrate on site the effective use of activated carbon,
establish a treatment protocol for use throughout the remedial program, and conduct pilot studies to optimize treatment. After EPA's departure, Calgon's modular carbon treatment unit continued treatment of leachate and surface
liquids on the southern third of the site. More extensive
data on activated carbon treatment have been collected, and
studies will continue to determine a permanent system to
deal with the collected leachate.[19]

Long-Term Decontamination and Rehabilitation

With the relocation of the residents in the immediate
areas, and construction of remedial measures for the critical
lower one-third of the site well on the way, development of
a demonstration grant was important to ensure the continuation of government action, with the goal of complete decontamination of the site and the return of impacted people to
their homes. Such a plan, which is now in effect, has the
following components:

(a) Complete the leachate control system. This
    includes design and construction of the leachate
    collection system, holding tanks, site prepara-
    tion, and the clay cover.

(b) Begin leachate treatment. The installation of a
    leachate collection system will require a treat-
    ment scheme to ensure contaminated liquid is
    properly treated and disposed of, both during
    trenching and on a more permanent basis.

(c) Continue environmental monitoring. Sampling and
    complex analysis of ambient air, soil, basement
    air, ground and surface waters and sumps will
    continue to define the extent and velocities of
    chemical migration, evaluate the existing and/or
    potential pollution of deeper aquifers, and dem-
    onstrate the effectiveness of remedial measures.

(d) Initiate additional remedial measures. Expanded
    and continued monitoring under (c) above may
    suggest the need to design and construct additional

remedial measures. For example, if monitoring
shows the deeper aquifer to be contaminated, measures such as well systems and grout curtains may
be needed to curb or prevent contamination. Also,
the environmental data may suggest a significant
extension of lateral migration well out into the
surrounding areas, possibly along old stream beds,
thus requiring local leachate control systems.

(e) Begin home rehabilitation. Some work will be
devoted to investigate techniques and materials
that can seal basements and the school sufficiently to make them safe and acceptable for future
use. Such measures should be effective for both
remedial and preventive purposes.

(f) Continue health studies. Epidemiological investigations and risk assessments will continue, with
both acute and chronic effect studies. Included
will be door-to-door surveys throughout the entire
perimeter of the site, complete blood counts, a
review of cancer cases, hospital surveys, compilation of data from case records, and finally,
tabulation and statistical analysis, including,
if possible, of appropriate relative risks.

(g) Land use studies. Complete decontamination at
Love Canal will make possible the reuse of the site.
Therefore, land-use alternatives will be evaluated
on health, social, and economic grounds, and a
possible rehabilitation program will be developed.
Some initial thoughts would include the reestablish-
ment of the residential area, development of a park
complex, a creation of a community service complex.

Plans for long-term decontamination and continued health
studies have been recently further spurred by the President's
declaration of a health emergency at Love Canal on May 21,
1980. The President's order, which supercedes the August 7,
1978 emergency declaration, is likely to lead to further precautions. One action may be evacuation of the remaining 710
families residing at Love Canal in rings 2 and 3.

Conclusion

Love Canal is a vivid example of the extremely serious
environmental problems that exist because of improper disposal of hazardous wastes. Other ticking time bombs may be
found throughout the nation, and this gives Love Canal a pre-

cedent-setting significance. For example, according to EPA estimates, 80-90% of the 30 to 40 million metric tons of hazardous waste produced each year are being disposed of in ways that will not meet forthcoming standards under the Resources Conservation and Recovery Act.[20]

In the future we can expect to find outselves frequently with insufficient information to guide our actions. For example, hard data on the types and amounts of waste dumped at various sites will be unavailable. Environmental data (air, soil, and ground-water) may have to be developed to begin assessment of risk to public health and environment. In this context it should be clear that resources and equipment needs for sampling and analysis may be large, especially if we recognize that at Love Canal the cost of federal, state, and local monitoring alone has exceeded $1 million.[21] It is therefore an important priority to reduce this resource load through the development of key environmental and health indicators.

From the experience at Love Canal, it is clear that the remedial measures must be applied on a comprehensive basis. Solutions will require innovative thinking in the application of standard practices as well as development of new technology. A particular challenge is solving the problem of ground-water contamination. Once remedial systems, such as leachate collection plants, are constructed, it is essential to maintain them, not only for the sake of long-term protection of health, but also for protection of initial investments in remedial technology. Funding will be a problem. Decontamination of Love Canal may ultimately carry a price tag of more than $20 million, not counting costs of evacuation and relocation.[22] While funding at that level is probably forthcoming for Love Canal, finding appropriate funding for smaller, less dramatized sites will be difficult.

Aside from obtaining enough useful information and adequate funding for cleanup, a key problem for the future will be the regulatory authority of government. EPA's authority to clean up sites is often quite limited. The Resource Conservation and Recovery Act (RCRA) will not enhance this authority.[23] Many states, however, will have broader authorities. To make maximum use of learning that takes place at various locations in the country, it is important to coordinate efforts of state authorities as much as possible. Here, the Federal Government can help by providing assistance to the states.

Finally, a most important aspect of the future is dealing with the legitimate concerns and fears of affected people. If Love Canal is any guide, many hazardous waste situations will heighten public concern, and in some cases, provoke extreme anxiety. Government has the responsibility to inform the public adequately and to involve the public in the process of monitoring and cleanup.

References and Notes

1. For representative accounts of William Love's endeavors, see M. Brown, *Laying Waste: The Poisoning of America by Toxic Chemicals* (Pantheon Books, New York, 1979), pp. 7-8; E.C. Beck, "The Love Canal Tragedy," *EPA Journal*, 5 (January, 1979), 16; New York (State), Dept. of Health, Office of Public Health, *Love Canal: Public Health Time Bomb* (Dept. of Health, Albany, 1978), pp. 2-3 (hereafter cited as *Time Bomb*).

2. *Time Bomb*, p. 131; New York (State), Dept. of Health, *Supplemental Order in the Matter of the Love Canal Landfill Site Located in the City of Niagara Falls, Niagara County*, Section IV, Appendix A.

3. *Time Bomb*, p. 108.

4. R.P. Leonard, P.H. Wirthman, and R.C. Ziegler, *Characterization and Abatement of Groundwater Pollution from Love Canal Chemical Landfill, Niagara Falls, New York* (Calspan Report, ND-6097-M-1; Calspan Corporation, Buffalo, New York, August, 1977).

5. New York (State), Dept. of Health, Office of Public Health files.

6. *Time Bomb*, pp. 6-7.

7. *Time Bomb*, p. 6. For a chronology of events, see *Time Bomb*, pp. 23-24.

8. *Time Bomb*, p. 8.

9. *Time Bomb*, pp. 13-16.

10. D. Picciano et al., *Pilot Cytogenetic Study of the Residents of Love Canal, New York* (Biogenics Corporation, Houston, Texas, 1980).

11. G.B. Kolata, in "Love Canal: False Alarm Caused by Botched Study," *Science*, 208 (13 June 1980), 1239-1242, summarizes the scientific dissatisfaction with Picciano's conclusions. Letters including one by Picciano himself,

disputing Kolata's account appear in *Science*, 209 (15 August 1980), 751-756.

12. New York (State), Dept. of Health, *Order in the Matter of the Love Canal Chemical Waste Landfill Site Located in the City of Niagara Falls, Niagara County* (Albany, 2 August 1978).

13. Hugh Carey (Governor), telegram dated 3 August 1978 to President Carter requesting federal assistance under P.L. 93-288.

14. Conestoga-Rovers and Associates, *Phase I: Pollution Abatement Plan - Upper Groundwater Regime* (Conestoga-Rovers and Associates, Waterloo, Ontario, 13 June 1978).

15. Fred C. Hart Associates, *Preliminary Assessment of Cleanup Costs for National Hazardous Waste Problems* (Environmental Protection Agency, Washington, 23 February 1979).

16. *Time Bomb*, pp. 19-21.

17. W.J. McDougall, R.A. Fusco, and W.P. O'Brien, "Containment and Treatment of the Love Canal Landfill Leachate," paper presented at the Water Pollution Control Federation Conference, Houston, Texas (11 October 1979).

18. The "Blue Magoo" system is described in M.K. Gupta, *Development of a Mobile Treatment System for Handling Spilled Hazardous Materials: Final Report* (EPA 600-2-76-109; Environmental Protection Agency, Washington, July, 1976).

19. F. Freestone (EPA Office of Research and Development, Edison, New Jersey), Memorandum to Fred Rubel (Chief, Emergency Response Branch, EPA Region II), 20 November 1978.

20. *Federal Register*, 43 (18 December 1978), 58947.

21. Information contained in EPA files.

22. Information contained in EPA files.

23. For a discussion of the provisions and deficiencies of RCRA, see U.S. Congress, House, Committee on Interstate and Foreign Commerce, Subcommittee on Oversight and Investigations, *Hazardous Waste Disposal: Report...*, 96th Cong., 1st sess. (Committee print 96-IFC 31; September 1979), pp. 28-29, 47-50.

Lois R. Ember

7. Love Canal: Uncertain Science, Politics, and Law

Introduction

Love Canal is the nation's first notorious episode of land pollution. This violated chemical grave haunts the residents of the area, stalks the company that buried the wastes, and tests the mettle of governments charged with protecting their citizens and the environment. In short, the precedent-setting potential of Love Canal has hampered efforts toward a permanent solution to the cleanup of the landfill, the speedy relocation of residents, and the execution of well-designed health and environmental studies.

As the first chemical dump to release its toxic legacy, it is fitting that Love Canal is located in that part of New York state known as the Niagara Frontier. Resolution of the not-so-mute medical, legal, and moral issues swirling about Love Canal will extend the frontiers of environmental health, the law, and corporate responsibility.

Testing the frontiers of events has always been Love Canal's hallmark. William T. Love, a turn-of-the-century entrepreneur, tested his dream there. He would build a navigable power canal connecting the upper and lower levels of the Niagara River, and this 7-mile-long trough producing cheap electricity would bring industry and commerce to his planned model city. His dream, however, never materialized. An economic depression and the invention of alternating current contrived to stifle it after only 3000 feet of his canal had been dug.

Adapted by permission from an article in the 11 August 1980 issue of Chemical and Engineering News, a publication of the American Chemical Society.

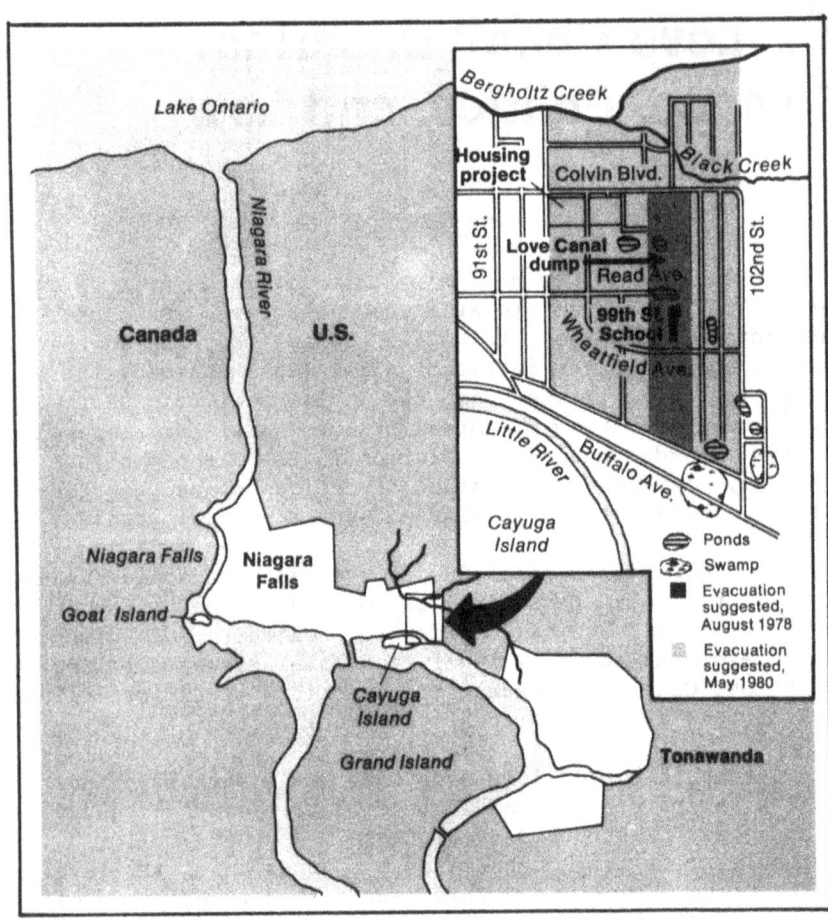

Figure 7.1.   Map of Love Canal area indicating widening
area of evacuation.

The abandoned canal first served as a neighborhood swimming hole and later became a dump. In 1942, the Hooker Electrochemical Corporation (now Hooker Chemicals & Plastics Corporation), a subsidiary of Occidental Petroleum Corporation, obtained permission to dispose of its chemical wastes in the canal. In 1947 it bought the 16-acre property. And for 11 years until 1953, when it deeded the land to the Niagara Falls Board of Education for a token $1.00, the company used Love Canal as a burial vault for more than 21,000 tons of its wastes.

It wasn't a final resting place for this chemical soup. Twenty-five years later, excessive rains and melting snow penetrated the canal's cover and the rising water table carried the "Love Canal leachate," a sweetly aromatic, oily black ooze, out of its grave and into the backyards and basements of homes that had been built adjacent to the landfill.

The canal residents believed that the chemical ooze brought disease and death, and they lobbied for relief. As Librizzi (Chapter 6) has noted, nearly two years of study by the EPA and the New York Department of Public Health led by mid-1978 to the identification of more than 80 chemical compounds, eleven of which were suspected animal carcinogens, and one compound, benzene, was a proven human carcinogen. This finally brought initial relief to the residents. Some 239 families in the first two rings of homes bordering the canal were evacuated, their homes were purchased by the state, the 99th St. School built over the center of the landfill was closed, and a six-block area was cordoned off by an 8-foot-high chain link fence. (See map, Figure 7.1.)

Since then additional environmental and medical studies, some under dispute, have tracked chemical migration and disease beyond the 1978 evacuation lines. Attendant publicity has made the homes caught in the ever widening net worthless, and area citizen groups, of which the Love Canal Homeowners Association is the most vocal and effective, have been clamoring for permanent relocation. But caught in bureaucratic cross fire, the residents remain in economic, medical, and legal limbo. Migrating toxic chemicals plus government ineptness have not brought them evacuation, only fear, frustration, and a nightmare beyond their control.

After President Carter's second declaration of a health emergency on May 21, 1980, (an earlier one came in August 1978) the only humanitarian gesture remaining seems to be the evacuation of the remaining 710 families in the area.

This is the action being urged by Representative John J. LaFalce, whose district includes the Love Canal area, and New York Governor Hugh Carey. Cynical but realistic, the residents expect to be moved out in this Presidential election year. At this writing, in September 1980, state and federal officials are still negotiating the financing of the move, but there seems little doubt that evacuation will have taken place by the time this volume is published.

Fixing the Blame:
Is Industry Responsible?

One of the most vexing questions at Love Canal is who is ultimately responsible for creating the dangerous situation that exists today. Hooker, the company that buried 21,800 tons of chemicals at Love Canal (see Table 7.1) disavows all liability. It argues that it disposed of the chemicals in an impervious clay vault using state-of-the-art techniques for the period. Hooker also says that under threat of eminent domain, it transferred the property to the school board, which wanted to build a school on it, and that the deed of conveyance contained a paragraph warning that the landfill had been filled with "waste products resulting from the manufacturing of chemicals," and that Hooker disclaimed all liability for injuries occurring as a

Table 7.1. Chemicals disposed by Hooker at Love Canal.

Type of Waste	Estimated tonnage
Miscellaneous acid chlorides	400
Thionyl chlorids	500
Miscellaneous chlorinations	1,000
Dodecyl mercaplans (DDM)	2,400
Trichlorophenol (TCP)	200
Benzoyl chloride	800
Metal chlorides	400
Liquid disulfides (LDS/MCT)	700
Hexachlorocyclohexane (BHC, Lindane)	6,900
Chlorobenzenes	2,000
Benzyl chlorides	2,400
Sulfides	2,100
Miscellaneous 10% of above	2,000
TOTAL	21,800

Source: Hooker Chemicals and Plastics Corporation

result of these "industrial wastes" after the property transfer. Furthermore, on subsequent occasions it went before
the school board to warn it against construction on the property that might result in disturbance of the "dangerous
chemicals" buried there.

Additionally, Hooker claims that before property transfer it placed over the landfill a clay covering whose in-
tegrity was later breached by the construction of city roads,
an expressway, and the removal of thousands of cubic yards
of dirt. By Hooker's analogy, precipitation penetrated
the violated cover, and the buried chemicals rose with the
water table, overflowing the canal banks like a stoppered,
unattended bathtub.

Hooker will have to defend these claims in court. Com-
pensatory and punitive damage suits brought against the company by New York state, the Justice Department on behalf of
EPA, and 1300 private individuals total billions of dollars,
with some estimates as high as $11 billion. Sophisticated
analytical tools are dissecting the chemical miasma of Love
Canal, but the courts will resolve the conflicting responsibilities and liabilities born of that miasma.

No public documents have yet surfaced to support Hooker's claim of a clay cover, although Hooker may have such
documents, which it will bring out in court. Frank Ventry,
an operating engineer for the city of Niagara Falls who
worked at the landfill site after Hooker deeded it to the
school board, says, however, that it was the city that
covered to grade the full length of the canal, including
the two ends that Hooker claims to have covered. He says
this was done mostly by backfilling from the mounds of dirt
piled on the banks of the canal from the original excavation. This point will be argued in court because it bears
on the question of whether Hooker properly secured the
landfill before it turned it over to the school board.

Although the canal has been deemed an exceptionally
fit site for hazardous waste disposal by Hooker, there are
some natural geologic formations such as swales--dried up
streambeds--and a sand lens--a thin layer of sand inter-
spersed between more impervious clay layers--that may have
served as preferential routes of migration for the chemicals.

It is not reasonable to assume that Hooker or any company dumping in a landfill in the 1940s and 1950s would
have been aware of such geologic formations, but it might

be reasonable to assume that diligent continuous monitoring would have picked up wayward chemical migrations via these formations. The courts will consider the monitoring issue, especially who can be held liable for the lack of monitoring.

Certainly, one of the major routes of chemical migration was, to use Hooker's analogy, the overfilled bathtub. That is, the rising water table in the clay bathtub carried leachate upward and laterally outward through the more permeable top 5 feet of soil.

But there were two other manmade conduits, according to the Love Canal Homeowners Association consultant, Stephen Lester, that were unearthed during remedial construction in 1978-79. One was a French drain, put in by the school board, that completely encircled the 99th St. school. The drain funneled water and chemical leachate away from the school to storm sewers on 99th St. whose contents flow to the Niagara River, south of the canal. Another was an illegal catch basin put in by a home owner on 97th St. that carried water and leachate to storm sewers on 97th St. whose contents eventually end up in Black Creek, north of the canal. An internal EPA memorandum notes that chemicals buried at Love Canal, including dioxin, have been found in the sediments and water of Black Creek and the Niagara River. The storm sewers may be how the chemicals got to the creek.

According to state geologists who made hundreds of borings around the landfill during the construction of the leachate collection system, the soil strata of the landfill are fairly well defined. From ground level to about 5 feet below grade, the soil is sand and silt--a permeable layer that allows leachate from the canal to pass easily. Below this layer to a depth of about 25 feet are lacustrine clays, which are very impermeable. Below this layer, to a depth of 40 feet, is glacial till, a highly impermeable layer, and below the glacial till is limestone bedrock.

The geologists do not know the exact depth to which the canal was excavated, but seismic surveys over its central and northern portions suggest that the excavation was 12 to 13 feet deep. The excavated depth of the southern section is not known, and therefore the thickness of the protective clay layer between buried chemicals and bedrock is not known. Hooker did most of its dumping on the southern section of the canal, and the highest levels of contamination and reported illnesses were found in homes abutting this portion.

Despite these geological uncertainties, Frank Rovers of Conestoga-Rovers, a consultant to the City of Niagara Falls and a designer of remedial measures at Love Canal, supports Hooker's contention that the nature of soils at Love Canal made it an ideal burial vault for chemicals. He asserts that "the hydrogeological environment of Love Canal would allow for adequate design of a secure disposal facility," even under today's more stringent Resource Conservation and Recovery Act regulations. The problems at Love Canal, he says, resulted from a failure to monitor and maintain the site after closure, and from disruption of the closure by subsequent construction.

The courts, of course, will decide who was liable for monitoring and maintenance: Hooker, which no longer owned the land, but which knew the nature of the chemicals it buried there, and had the expertise to monitor and analyze for migrating chemicals; or the board of education and the city of Niagara Falls, which owned the land and developed it, but which might not have fully comprehended the nature of the chemicals disposed by Hooker, nor had the expertise and perhaps even the understanding to acquire outside expertise to monitor the site. Or the courts may find all liable.

Monitoring the Damage: What's in the Sludge?

For those unfortunate enough to live near Love Canal, more important than the degree of Hooker Chemical's responsibility is the question of what harm may come from prevailing environmental conditions. One approach to estimating the health hazard has been to monitor the chemicals present. Historically, this was the first approach used, and by now it has been carried to great lengths. For example, the state health department has launched a massive environmental monitoring and analysis effort including 5924 soil samples from more than 650 homes and the landfill, of which 2500 have been analyzed, and more than 700 air, sump, and water samples and analyses.

More than 400 separate chemicals have been identified by the state and EPA, including the 2,3,7,8-isomer of dioxin--one of the most potent toxins known to man. This finding was not unexpected. Hooker disposed of trichlorophenol at the canal, and dioxin is the unintentional contaminant of trichlorophenol. But what is alarming is where it was detected: in a composite of surface landfill soil at 5.3 ppb; in a subsurface soil sample taken from the backyard of a 97th St. home abutting the canal at 6.7 ppb;

in sediment samples from Black Creek, which flows into the Niagara River, at 30 ppb; and in a composite sample of crayfish taken from Black Creek at 3.1 ppb.

According to health commissioner Axelrod, the state's environmental sampling program confirmed the presence of a "host of toxic chemicals in the landfill"; documented the outward migration of some of these chemicals, "principally through porous topsoil and drainage pipes"; and confirmed "diminishing degrees of soil and air contamination as one moves further away from the landfill." EPA's studies, described in detail by Librizzi (Chapter 6) generally confirm the state's findings.

Whether chemicals have migrated vertically downward through bedrock to underground aquifers has not yet been established. The state Department of Environmental Conservation sank three wells to bedrock around the canal, sampled from these twice, and found no organic chemical contamination of the water. But the number of wells is too small to be able to rule out contamination. Therefore, the agency is now sinking seven to 10 additional wells around the canal, and EPA--if it is permitted--will sink from 100 to 200 monitoring wells in a grid pattern around the canal and throughout the community.

Some of EPA's wells will go to bedrock to monitor for aquifer contamination and to determine the hydraulic flow in the area. As planned, the wells will determine the efficacy of the remedial construction effort and map the present flow of chemicals through the community.

At this time, near the end of 1980, the state's environmental efforts have wound down, though ground-water sampling from deep wells will continue. At the same time, EPA is continuing in high gear. In August 1980, EPA launched a major environmental monitoring effort. The effort will include collection and analyses of a massive amount of indoor and ambient air samples, soil, ground water, surface water and sediments, drinking water, and sump water samples. EPA's Thomas Hauser is overseeing the $5.4 million integrated assessment project, which is being managed at Love Canal by GCA Corp. of Bedford, Mass. The effort is expected to be completed in February 1981, at which time EPA will be able to estimate the risk to residents of living near the landfill. By that time, however, that estimate may be moot. By then a political decision already may have been made to move the residents out.

Ironic as it may sound to complete risk estimates after the potential victims are evacuated, it must be remembered that EPA's purpose was initially to monitor the site and to establish culpability. A summary of EPA environmental data and its bearing on the lawsuit against Hooker Chemical is given in Table 7.2.

In addition, the environmental data collected by the state and the EPA was responsible for "remedial action" that is now completed. This action consisted of a tile drainage system's being constructed by Conestoga-Rovers under contract to city of Niagara Falls. The system, which was completed in January, 1979, is described in detail by Librizzi (Chapter 6).

Direct Studies of Health:
On-Site Epidemiology

Beyond monitoring the environment for chemicals and estimating risk based on existing toxic effect data, a second and more direct approach is to search for harm directly among the affected populations. By this time, there has been a wide range of such studies, conducted by the State, the EPA, and various nongovernment scientists. Together with the ongoing litigation, government haggling and bungling, these studies have roused the anxiety of the residents, contributed to the destruction of families, and rent the social fabric of the community.

One of the first health effects studies led to the finding by the State that miscarriages at Love Canal were 1.5 times the norm, and that birth defects clustered around the southern portion of the Canal. This study, completed in summer 1978, led directly to the first declaration of a health emergency by State Commissioner of Health Robert P. Whalen in August 1978 and prompted the evacuation of the first group of 239 families soon thereafter.

A separate epidemiological study was launched by the Love Canal Homeowners Association under the guidance of its scientific adviser, Beverly Paigen, a cancer research scientist at Roswell Memorial Park Institute in Buffalo, New York. Paigen apparently was able to correlate clusters of diseases--miscarriages, birth defects, respiratory illnesses, and others--with swale or wet areas. Swales, several of which cross the canal, are dried up streambeds that may serve as preferential routes of chemical migration from the canal.

Table 7.2. Summary of EPA environmental data for lawsuit against Hooker.

Source	Date Collected	Chemicals detected	Comments
Water from storm sewers all three locations	August 1978	Hexachlorocyclohexane (1.4-50 ppb), trichlorophenol (0.1-11.3 ppb), chlorobenzene (0.3-0.4 ppb), tetrachlorobenzene (18-130 ppb), & others	Exceed EPA's water quality criteria for human exposure
Sediment samples from 10 locations in Black Creek, Bergholtz Creek, Cayuga Creek, Little River, Niagara River	Sept. 1978	Benzene (0.1-1.7 ppb), hexachlorocyclohexane (31-1200 ppb), trichlorophenols (0.5-90 ppb), chlorobenzene (0.4-2.9 ppb), tetrachlorobenzene (24-240 ppb)	Exceed EPA's water quality criteria for human exposure. Chemicals also found upstream but at substantially lower levels
Drinking water from home taps	Sept. 1978	Chloroform (25 ppb), bromodichloromethane (6 ppb), tetrachlorobenzene (2 ppb)	No concurrent sampling done at drinking water plant, so precise source of chemicals not determined; may have resulted from water treatment
Ambient & basement air	Sept. 1978	Benzene, tetrachloroethylene, trichloroethylene, dichlorobenzene in the moderate-ppb range	The incremental cancer risk was calculated as 1 in 100 to 1 in 1000 for some of the homes tested

Table 7.2, continued

Source	Date Collected	Chemicals detected	Comments
Leachate from leachate collection system (southern section)	June 1979	Lindane (γBHC) (160 ppb), dichlorobenzenes (7200 ppb), & others	
Leachate from French drain system around 99th St. School	June 1979	Lindane and dichlorobenzenes & others at one-tenth concentration of the same chemicals found in landfill leachate system	
Surface water & home sump pumps (northern section)	June 1979	Phenol (0.4–4.7 ppb), pentachlorophenol (54–70 ppb), trifluoro m-cresol (12–102 ppb), [trichlorobenzene (1030 ppb), hexachlorobenzene (250 ppb) found in sumps only]	Exceed water quality criteria for human exposure
Ambient & basement air	Nov. 1979	Light to moderate contamination by a number of chemicals, including benzene	Only 2 homes sampled; location not listed; 1 in 10,000 additional risk of cancer

Source: United States Environmental Protection Agency

Her survey, not without flaws, was nevertheless found
by a then-Health Education and Welfare panel to merit further study and has, in part, been validated by an expanded
state epidemiological survey. The state's study also correlated swales and other historic wet areas with higher
rates of miscarriages, lowered birth weights and, possibly,
elevated rates of birth defects.

Raw data from this epidemiological study have not been
made public; the state argues patient confidentiality. EPA,
Hooker, the home owners association, and others have all
sought the data. A U.S. District Court in Buffalo is now
mediating release of the data to EPA, Hooker, the home owners, and also to the Center for Disease Control, which began
its health studies toward the latter part of 1980.

Lois Gibbs, president of the home owners association,
believes that the state has withheld the data for three reasons. First, she says, there is a high health risk to residents living in the area. The state knows this, but cannot
afford to spend much more than the $35 million it already
has spent on remedial cleanup and purchase of 239 homes.
Second, if the state acts to relocate permanently all affected Love Canal families, a precedent will be set. With its
own recent survey identifying 680 hazardous waste dumps
across the state, 68 in Niagara County alone, the state looks
askance at this prospect.

And finally, Gibbs claims the state designed a flawed
questionnaire, collected information badly, and interpreted
the data poorly.

"If that study were to go out into a scientific forum,
it would be ripped apart," she says. It wasn't exactly
"ripped apart," but it was recently rejected by Science magazine for, in part, not identifying which of the two control
groups it used was more appropriate for comparison with the
exposed population.

Health studies went into abeyance until May 1980 when
a "leaked" story appeared on the television evening news,
announcing the unconfirmed results of an uncontrolled
"study." The study, conducted for EPA by Dante Picciano,
scientific director of Biogenics Inc., a small Houston firm,
found broken or extra chromosomal (genetic) material in 11
of 36 residents, which it linked to chemical exposure, and
which it said foreshadowed miscarriages, cancers, and other
disorders.

Preposterous though it may seem, an attempt was made to gloss this limited, uncontrolled study, undertaken to gather legal evidence for EPA's lawsuit against Hooker, with a scientific sheen by having two expert panels review it. These panels assailed the study for lacking controls and for other methodological deficiencies after reviewing, not primary chromosome preparations, but the report itself and photocopies of photographs of these preparations. Still a third panel of academicians, reviewing the primary material, supported some of Picciano's findings and called for an expanded, controlled chromosome study of residents.

Paigen selected the 36 residents for the chromosome study, and she says that she also had controls picked out, but that she and Picciano were rebuffed by EPA on this issue.

Paigen also was involved in another study that surfaced with the chromosome test. This one, designed and conducted by Stephen Barron, assistant professor of neurology at the State University of New York, Buffalo, charted the conduction of the nervous impulse along peripheral nerves in 35 residents presumably exposed to the neurotoxins found at Love Canal, and in 20 control subjects.

Barron, a neurologist not a statistician, at first thought his data showed a slowing of conduction in exposed residents. At the request of EPA, he submitted his data to rigorous statistical analyses, and in the process "the difference in the mean nerve impulse velocities of residents vs. controls lost power, but the direction of change was still there: a slowing of nerve conduction in Love Canal residents," Barron explained.

The velocity of conduction is a sensitive indicator of neurological damage, appearing long before such overt symptoms as tingling and numbness occur. Nerve conduction, therefore, may be an excellent biological marker for toxic exposure. Barron is encouraged enough by his findings to seek funds for a larger study from EPA and private foundations.

Although Barron's preliminary nerve study was not statistically significant, but nonetheless "meaningful," and Picciano's limited study of chromosomes was widely criticized as no better than a legal fishing expedition, they became the scientific veneer to the federal government's decision to declare--for the second time in two years--a state of emergency at Love Canal. The government had little choice. The incredibly poor manner in which the chromosome data were

handled raised the already high level of fear and hysteria
still higher. This fear was little quelled by the government's insistence on temporary relocation of the 710 affected
families, so chosen because they lived in the tax abatement
area created by the state legislature to encourage people
to remain in the area.

On May 21, the day the emergency declaration was an-
nounced, Governor Carey called Love Canal a "neighborhood
of fear" and said that temporary removal was not sufficient
to alleviate "the disruption of their lives and the continued
trauma." He called on the federal government to match the
state's efforts by purchasing the homes that are now "presumed to be contaminated."

As the tug-of-war between the state and the federal
government continues, the parade of health studies marches
on. In June, the Environmental Defense Fund, with funding
from private foundations, began preliminary tests that will
chart growth and maturation of children under 18, and nerve
conduction in children and adults. Joseph Highland, head
of the environmental group's toxic chemicals program, is
spearheading this effort. Paigen is an unpaid consultant on
both studies and Barron will conduct the nerve studies.

The assumption underlying both studies is that simple,
noninvasive, biological markers of health effects related
to toxic chemical exposure exist. Growth and maturation
and nerve conduction may be such markers. Growth and development are retarded in animals exposed to toxic chemicals;
children also may be so affected. The preliminary study,
which the scientists feel is adequately controlled, will
test this hypothesis. Because it was not a government-
sponsored study, resident participation, including children
who already had moved from the area, was high--about 60 to
70%. The collected growth data are now being analyzed, and
the nerve-conduction tests are just getting under way.

Highland has submitted a proposal to EPA for funding
to expand the growth and nerve conduction studies. His proposal, though well received, has not yet been funded by
the agency. However, he recently received $50,000 from the
Frank S. Gannett Newspaper Foundation to support these
efforts.

The immediacy of testing and the high rate of participation may give added weight to Highland's preliminary (and
now expanded) efforts, especially if continuing Federal
health studies at the Center for Disease Control (CDC) are
delayed further and permanent relocation is begun, with the

effect of scattering still more residents and lowering the participation rate.

CDC's studies, which were to begin the first week in August, originally were to be conducted by EPA. Sometime between the emergency order of May 21 and early June, this responsibility for health testing shifted to CDC. The order came from the White House, possibly in response to EPA's poor handling of the chromosome study, possibly because the Office of Management and Budget and the Office of Science and Technology Policy believe that "no program to establish an emergency capability in health effects should be developed by EPA," or possibly because New York health commissioner David Axelrod, who, it is said, bears little love for EPA, asked that CDC assume responsibility.

At any rate, three months after the emergency declaration, federal health testing, to be headed by CDC's Clark Heath, had not begun. CDC proposed to begin what it calls its first-level tests--"executive" physical examinations and routine blood tests--on all Love Canal residents. From this larger group, individuals selected at random would be included in the so-called second-level specialized tests, whose results would be compared with two control groups.

The second-level tests would scan for toxic effects on chromosomes, nerve conduction, the reproductive system, the immune system, growth and development, and psychological well-being. About 40 subjects for the chromosome study were chosen from 12 families whose homes were the most highly contaminated by chemicals.

According to CDC's Renate Kimbrough, physical examinations were to have been completed by the end of 1980 and the specialized tests were to have been completed but not analyzed by December 1980. This schedule has been thrown out of kilter by the residents' boycott. CDC, attempting to balance immediacy against sound methodology, wants to select subjects for its specialized tests. Six community organizations want all residents to participate.

CDC claims that there is not enough money or laboratory resources to test all residents for chromosome damage, to take just one example. Furthermore, Heath claims that though the residents are concerned about their health, their more immediate concerns are "relocation, and economic dislocation and attendant uncertainties." But, he says, "the health studies shouldn't be looked upon as a source of any useful data in the near future for deciding the relocation issue."

Disregarding the relocation issue, Heath admits that
chemical toxicology, an embryonic science, places a constraint on the types of studies that now can be devised to
show a cause and effect relationship between exposure to
toxic chemicals and disease.

Heath says: "It is a difficult problem to say a low-
level exposure to this or that chemical, let alone to these
chemicals in combination, does or does not cause illness.
Most of the chemicals cited at Love Canal appear to be at
low levels. We know from traditional acute toxicological
principles that these aren't the kinds of levels one would
expect to produce outright acute toxicity. But the concerns
here deal with unknown and rare effects. That, plus the
fact that we are dealing with chemicals acting in combinations that have never been tested, makes it difficult to
predict even which effects should be looked for."

CDC's tests, if they are ever begun, are not likely to
prove or disprove a causal connection between exposure to
chemical contamination and disease among the residents of
Love Canal. Because of the time lapse--two years for the
families removed from chemical contamination in August 1978--
probably the only valid study that now can be conducted,
says Paigen, "is a registry of all current and former residents of Love Canal with health followups over the years."

At present, in September 1980, the best that can be
said about health effects among the 2618 people who once
lived in the Love Canal neighborhood is drawn from the
State's effort during 1978 and 1979. The state drew 4,386
blood samples from 3,919 people (some controls) and conducted 114,036 separate blood tests. It conducted 11,138
field interviews with Love Canal and surrounding area residents; it is searching for and conducting health followups
on some 2000 people who once lived in the area but moved
before 1978; and it conducted 411 physical examinations of
workers who performed remedial construction work at the
landfill.

From the health and environmental data it has amassed,
the state health department concludes that:

● There is an excessive rate of miscarriages and
  low-birth weight babies and, possibly, congenital
  birth defects in certain sections of the Love
  Canal area, although these cannot be related to
  chemical exposure.

- There is no evidence for an excess of liver and kidney abnormalities, leukemia, and other blood diseases.

- Cancer rates are within normal limits for this population.

- Computer analysis of the epidemiological survey produces no evidence of "unusual patterns of illness."

Thus, the health department has not been able to correlate the "geographic distributions of adverse pregnancy outcomes" with evidence of chemical exposure. "At present, there is no direct evidence of a cause-effect relationship with chemicals from the canal." The caveat, of course, is that there still might be a cause-effect relationship that has not yet been identified.

Complicating the search for cause and effect is the fact that mounting strain may account for many of the illnesses reported by the residents. Unfortunately, it is not yet possible to separate chemical-induced illness from stress-induced disease. In the absence of sound medical science, anecdotal evidence fills the void. This evidence fuels the residents' fears and further adds to psychological stress.

The Politics and Law of
Uncertain Health Effects

The inconclusive yet fear-inducing nature of the health effects puzzle at Love Canal naturally lends itself to claims and counter claims that serve as much the political and legal interests of the parties involved as the clarification of uncertain science. In this respect Love Canal is no different from a number of other cases discussed elsewhere in this volume (Coates, Chapter 3), and the similarity to the nuclear accident at Three Mile Island is particularly striking.

In the view of Hooker's President Donald L. Baeder, Commissioner of Health Robert P. Whalen's August 2, 1978 declaration of a state of emergency "created a spark of fear in these people that blew up the Love Canal that we know today."

For instance, Baeder argues, "the concentration of chemicals that were detected in the basements...were not compared with any controls, nor were they compared with...

Occupational Safety and Health Administration standards on
allowable limits for these chemicals in a continuous working
environment." Nothing was done by Whalen, Baeder says, "to
try to put these into perspective as to the levels of exposure, and the risk associated with those levels." Furthermore, Whalen failed to have the epidemiologic data reviewed
by competent, independent experts.

The levels of chemicals measured in the basements of
the homes, Baeder says, were in many cases "a small fraction
of a per cent of what OSHA would permit for a continuous
working environment. So if you want our position on the
relative risk of exposure to chemicals found, it would be
extremely small." Baeder goes as far as to say that the
"level of exposure from all of the measurements that have
been made at Love Canal precludes these chemicals at the
detected level from having an effect on the health of those
people. Now, that does not mean that those people aren't
suffering from asthma, cancer, leukemia, any other abnormalities they claim. But what the state says is that those
disease rates are normal for that population."

But the state, in the person of Stephen Kim, the scientist who analyzed those 1978 samples and who is now director of the state's Toxicology Institute, says: "The primary
motivation for Whalen's action was the contamination on the
surface of the landfill," to which the residents had open
access, not the indoor air samples. "The highest risk was
the kind of exposure--body burden--that a child might get
playing on the surface of the canal, then eating without
washing his or her hands." Further, Kim says, the state had
solid and very compelling evidence for Whalen's order, environmental and epidemiologic evidence that was reviewed by a blue-
ribbon panel of government, industry, and academic scientists.

Addressing the indoor air levels, Kim says that OSHA
standards--those that the agency has in the form of threshold limit values--are inappropriate guidelines for indoor
household air levels, because they are set for a healthy
working population, generally male, exposed for only a part
of a 24-hour day, five days a week, with a two-day respite
of nonexposure. "It's not continuous exposure," Kim argues,
"so we don't think it's necessarily appropriate for protecting susceptible segments of the population--the very
elderly, sick people, the fetus, and developing children,"
who may be exposed on a more continuous basis in the home.

In addition, the homes that were contaminated with
chemicals usually contained a mixture of compounds that
could act in an antagonistic fashion or in a synergistic

manner. If the action were synergistic, and no one knows that it was, induced health effects would be greatly enhanced, notes EPA's Hauser.

In short, Kim answers Baeder's charges by saying that internal and external controls were available to him, cancer risk assessments were calculated and have been made public, and the epidemiologic findings as well as the environmental data were reviewed by outside experts before Whalen made his announcement. The variation of the data, which correspond to certain geologic features, acted as internal controls. External controls included some data from basements in downtown Niagara Falls and the investigators' own homes in Albany. Kim also says that he attempted to eliminate confounding variables such as chemicals found in consumer products and auto exhaust fumes by using marker chemicals such as chlorotoluene and chlorobenzene, not found in consumer products. These latter chemicals were found in some of the homes and therefore could be expected to have emanated from chemicals buried in the canal, not from normal consumer products.

Still Baeder insists: "The facts are, and the state has so stated, that there is really no evidence to connect the chemicals of Love Canal with the ill effects that are reported by the residents. No information on that." Kim agrees in part. "Of the chemical indicators that we have measured so far," he says, "there is not a statistically significant correlation between those and the health effects that we have studied so far. But that does not mean that such a relationship will not be found, it only means we haven't found it yet.

"Love Canal is an important landmark in environmental health; finding a cause and effect relationship in this type of situation has never been done before. But our effort is continuing intensively," Kim says.

There is no question that society is on a learning curve with respect to environmental health issues. There also is no question that chemical-analytical probes are more advanced than medical probes. But all that is moot.

Love Canal, because of an incredible series of events, has moved beyond the scientific pale: Decisions affecting people's lives and corporate actions are being based on political expediency and liability law. If the Love Canal saga is allowed to play out, however, it may extend the frontiers of corporate responsibility as well as those of law and environmental health science.

At the moment, Hooker has chosen to follow a more legalistic path. Other companies might have responded differently. But the problem a corporation faces today, says Robert
Roland, president of the Chemical Manufacturers Association,
is "when and under what circumstances does (it) respond to
the equities of a situation vs. the legalities of a situation. When does the equity concern of what you think is
the correct thing to do take precedence over the legal concern?" It is a difficult balancing act to maintain, Roland
admits, and there is no denying that on the question of
equity, "there has been a definite change in mindset" in
the past few years, Roland says.

He elaborates: "A decision made in 1980 would by
necessity be different from a decision that was made in 1975.
I think that enough has happened during that period to alter
the response that a company would give to a Love Canal-type
situation. The legal concerns--personal liability, environmental damage, all of the rest--are no less grave today than
they were then, but the recognition of the terribly adverse
impact that the media can have, that inappropriate governmental response can have, that human expectations can have
on such a problem, demands a response that may be less legalistic and more humanistic."

Hooker essentially took a legalistic stance. Baeder
says, "I believe that the company was very responsible in
the way it handled those chemicals, in the closing of the
canal, in the transference to the city for the use intended,
and in warnings to the school board when we were apprised
of situations (proposed land development)."

The federal government and the state of New York apparently disagree. The Justice Department in December 1979
filed a $45 million suit against Hooker for waste disposal
at Love Canal, part of a $124.5 million suit against the
company for hazardous waste disposal at four dump sites in
Niagara Falls. New York state in April 1980 filed a $635
million suit against Hooker. Both suits seek damages for
the consequences of the dumping of hazardous wastes at Love
Canal. Hooker has answered only the federal complaint,
denying liability.

Should the federal claim go to trial, two prime issues
will be litigated: adequate warning and ultrahazardous
activity. The concept of ultrahazardous activity is well
established, but not for cases involving the disposal of
toxic materials. If the court rules the disposal of toxic
wastes an ultrahazardous activity and imposes on Hooker the
excessive duty of care and responsibility that has been

established for other activities deemed ultrahazardous, then the limits of liability law will have been enlarged.

The federal complaint states that Hooker did not give adequate warning. Barry Trilling, the Justice lawyer handling the case, says: "Hooker had knowledge of what was in the site, it had better knowledge than anybody about the dangers and hazards of the chemicals; it had the scientific capacity to know what may or may not happen; and it was in the best position to let people know what the problems were." That succinct statement may be the government's strongest argument in its suit against Hooker. There has been a flurry of activity at EPA and Justice to gather the preponderance of evidence needed to establish the government's claim.

Whether Hooker chooses, in Trilling's words, to test "the fires of adversary litigation," or settle the suits out of court--which some believe it will do to avoid further adverse publicity attendant to the discovery of evidence during a trial--is a litigation management problem that Hooker will have to resolve. Certainly, adverse publicity already has hurt the company, even in actions that it has taken that are environmentally "acceptable." For example, this fall, Hooker will begin operating a $94 million energy-from-waste project at its Niagara Falls plant. The project's financing took five years, and the private underwriters admit that the Love Canal incident probably added a half point to the financing cost.

The harsh glare of adverse publicity also has fallen on governmental response to the challenges of Love Canal. Bureaucratic bungling, interagency haggling, and bows to political expediency that in part have characterized this response have been spotlighted in the press.

Conclusion

The time it has taken to parcel out blame, responsibility, and costs, though adding to mounting stress, has done little to uncover a relationship between exposure to toxic chemicals known to be buried in the canal and illnesses reported by the residents. Finding the tools to make this causal nexus strains both medical and environmental sciences and probes the frontiers of environmental health. New York and EPA have made fitful stabs at it over the past two years, and even now EPA and the Center for Disease Control (CDC) stand ready to mount more comprehensive and, it is hoped, better designed environmental and health studies and eventual risk assessments. But these

risk assessments, if they come, will come too late to wield
a significant influence on events. The horse has fled the
barn. Now the best that can be salvaged from the planned
federal effort is the beginning of a record chronicling
the possible health effects on residents exposed to long-
term, low levels of a mixture of toxic chemicals. The proposed studies cannot and will not satisfy the residents'
desire to know their future health prospects, or those of
their children. The studies cannot and will not tell women
whether they can bear healthy children. Nor will the studies yield enough information on which decision-makers can
set good public policy in the near future. But they will
be a beginning.

Love Canal, quite simply, has been a learning experience, one in which traditional scientific tools and modes
of inquiry have been found wanting. As the governments have
worked their way through this learning maze, mistakes have
been made, and inadequate explanations of findings have
been given to the residents.

Government ineptitude, however, does not change the
fact that toxic chemicals were placed in the ground by
Hooker. And, although Hooker believed that its disposal
practices were safe, neither it nor subsequent owners moni-
tored the chemical grave in any regular fashion for decades.
The lack of monitoring allowed migrating chemicals to go
undetected for nearly 25 years.

The pity of Love Canal is that so many lives have been
so badly disrupted, and an unfortunate but exceptional
opportunity to expand our knowledge of chemical toxicity has
in part been frittered away.

Suggested Readings

Beck, E.C. "The Love Canal Tragedy," *EPA Journal,* 5 (January, 1979), 16-19.

Brown, Michael H. *Laying Waste: The Poisoning of America
by Toxic Substances* (Pantheon Books, New York, 1979).

Culliton, Barbara. "Continuing Confusion over Love Canal,"
Science, 209 (29 August 1980), 1002-1003.

Dickson, David. "Love Canal Continues to Fester as Scientists Bicker Over the Evidence," *Ambio,* 9 no. 5 (1980),
257-259.

"High Contamination Levels Found in Love Canal Hazardous
    Waste Site," *Chemical Regulation Reporter: Current
    Reports,* 3 no. 22 (31 August 1979), 890-891.

Holden, Constance. "Love Canal Residents Under Stress,"
    Science, 208 (13 June 1980), 1239-1242.

Levine, Adeline. *The Love Canal* (Lexington Books, Lexing-
    ton, MA., 1981).

New York (State). Department of Health. Office of Public
    Health. *Love Canal: Public Health Time Bomb* (Dept.
    of Health, Albany, 1978).

Omang, J. "Chemical Firms Sued for $124 Million," *Washing-
    ton Post,* 21 December 1979, p. A1.

"Those Disastrous Studies at Love Canal." Editorial, *New
    York Times,* 17 October 1980, p. A30.

U.S. Congress. House. Committee on Interstate and Foreign
    Commerce. Subcommittee on Oversight and Investiga-
    tions. *Hazardous Waste Disposal: Hearings...,* vol. 1
    (96th Cong., 1st sess.; Serial 96-48, 1979).

U.S. Congress. Senate. Committee on Environment and Public
    Works. Subcommittee on Environmental Pollution.
    Hazardous and Toxic Waste Disposal: Joint Hearings...,
    pt. 1 (96th Cong., 1st sess.; Serial 96-119, 1979).

Part 2

The Structure of Technological Risk

8. Overview: Causality, Perception, and Uncertainty in Technological Risk

Previous chapters have extracted from a discussion of some celebrated cases a number of generic problems to coping with technological hazards. Readers who seek a more systematic understanding of hazards and hazard management may find that analysis somewhat arbitrary. Thus the next three chapters explore the fundamental structure of technological hazards. In contrast to the contributions of Part 1, specific cases play a minor role, and the emphasis is on how risk and hazard may be defined, measured, and grouped for management.

Not that the generic approach to hazard management makes its debut in these chapters. Indeed, a number of U.S. government agencies and interagency groups have formulated guidelines for identifying, classifying, and regulating specific hazards and/or groups of hazards. The Environmental Protection Agency (EPA) has established criteria for regulating water pollutants,[1] carcinogens in air,[2] and pesticides.[3] The Occupational Safety and Health Administration (OSHA) has promulgated a standard for regulating potential occupational carcinogens.[4] A major concern of the Regulatory Council is "the process of establishing regulatory priorities within and among agencies."[5] The Interagency Regulatory Liaison Group (IRLG) has issued a compendium[6]--with updates[7] every six months--on toxic substances on which two or more agencies are contemplating regulatory action.

In 1980, the Toxic Substances Strategy Committee (TSSC) recommended:

> The IRLG agencies should continue to explore opportunities for use of generic approaches to the identification and assessment of risks presented by classes of chemical substances that pose similar

hazards because of structure, use, or other common characteristics.[8]

The contributions presented in Part 2 are certainly consistent with the foregoing recommendations.

As already noted in Chapter 1, the sheer number and variety of technological hazards are threatening to overwhelm society's effort to cope. One goal of generic approaches is therefore the identification of principles that can be applied to groups of similar hazards. In addition, managing hazards requires that scientific knowledge of hazard causation and perception be applied efficiently to a selection of the most serious threats. A second goal of generic approaches is therefore to develop methods that will help society to set priorities for managing hazards. Since priority setting is essentially a political decision process, we should not expect its resolution by resorting to simple formulae. In particular, we should not expect that the seriousness of a hazard will be quantified easily via "objective" measures of science. Instead, we should anticipate an intriguing and sometimes contradictory interplay of scientific and subjective issues. To be successful, generic approaches must provide characterizations that address this interplay.

The size of the problem. A first step in generic analysis is to define the size of the technological hazard problem. To this end, Hohenemser, Kasperson, and Kates discuss in Chapter 9 a series of estimates based on mortality. Even while recognizing that technology has been a principal contributor to increased life expectancy in the United States, they find that 20-30% of U.S. mortality can be associated with technological hazards and that the total cost of coping may be as high as 8-12% of the 1979 gross national product. Beyond mortality they note a number of other costs such as ecosystem disruption, effects on social environments, and certain economic costs not quantifiable as direct hazard consequences. Their analysis establishes the outset that the technological hazard problem is a substantial one.

The causal anatomy of hazard. To define the scientific structure of hazard, Hohenemser et al. visualize causal chains that run from human needs to human wants to choice of technology; and from there "downstream" to release of energy or material that threatens human health and further to eventual health consequences. The causal chains represent a convenient template for organizing scientific data about hazard causation. They also serve to display control

opportunities, to map regulatory effort, to realize feedback possibilities, and to classify patterns of societal
response.

To discern how hazards may be grouped according to
causal structure, steps in the causal chain are characterized by a series of physical variables. This leads to a
causal taxonomy according to which hazards are described
by five independent variables. These appreciably expand
on the traditional view of risk as probability of mortality
and morbidity.

Interestingly, the hazard concept of Hohenemser et al.
is in conflict with attempts in Part 3 of this volume to
define acceptable risk in terms of quantitative risk criteria. For, as shown in Chapter 9, many of the technologies that have extreme scores on various dimensions of causal structure, and are thus "highly hazardous," do not have
high expected annual mortality and would therefore score
quite low on "quantitative risk."

The cognitive anatomy of hazard. The manner in which
hazards are subjectively judged is explored by Slovic,
Fischhoff, and Lichtenstein in Chapter 10. After a review
of the biases that affect human judgment, the authors describe a series of experiments designed to define the cognitive processes that underlie lay judgments of hazards. Subjects were asked to evaluate activities and technologies
according to such qualities as newness, delay of consequences, voluntariness, or degree of knowledge. Taken together,
the data show that lay subjects possess a rich, qualitative
understanding of hazards which is reproducible and self-
consistent over a range of selected populations.

The authors also provide two different comparisons of
lay perceptions of risk and scientific estimates of risk.
The first comparison plots lay estimates of annual mortality
against scientific estimates. This juxtaposition shows
that, short of a number of glaring discrepancies involving
serious errors in estimation, lay people can provide the
correct "order" of riskiness. In the second comparison,
"risk" is not defined as annual mortality but left to each
subject's own interpretation. Here "perceived risk" does
not scale nearly so accurately with observed frequencies.
The perceived risk of nuclear power, for example, scales
as high as the risk of auto accidents. It appears likely
that the differences observed in the two comparisons point
toward an important insight: lay people, when asked in an
open-ended way, naturally view risk as a multivariate concept that consists not only of "expected mortality" but

several other qualities. If this is so, then attempts to
set quantitative risk standards in terms of expected mortality are in trouble, quite independent of the hazard structure analysis of Chapter 9.

Taken together, Chapters 9 and 10 imply that a multivariate description is required both for defining hazardousness in a scientific sense and for understanding the
structure of subjective judgments of hazards. What is not
yet clear is how the structure of perception may be related
to causal structure: that will require specifically designed investigations. Yet even at the present stage of
development, one may conclude that hazard management that
incorporates the full scientific and perceptual description
of hazards will not be a simple process.

Hazard management under conditions of uncertainty.
What happens when science cannot define adequately the
causal structure of hazards as envisioned in Chapter 9?
To this end, Gori addresses in Chapter 11 the best example
of this condition--the case of carcinogenic hazard. Gori
spends considerable effort in reviewing just why carcinogenic hazard has a poorly defined causal structure; for
example, he cites many fundamental reasons why human risk
quantification based on animal experiments is bound to fail.
He then turns to the problem of risk regulation and proposes his "alternative scenario"--a qualitative taxonomy
of hazards based on grouping of materials or technologies
according to uses or benefits. Gori envisions making relative risk comparisons within groups in order to determine
relative safety:

> The standard-of-need agent for each class would natur-
> ally become its standard of safety or tolerable hazard,
> and new agents aspiring to be classified for the
> same use would be compared for safety with that
> standard. For example, sucrose would be selected as
> the reference for sweeteners, because it is the most
> widely used substance for sweetening, and because
> a long record of chronic exposure in mankind sug-
> gests side effects that are either ignored or ac-
> cepted as tolerable to the vast majority of users.
> The safety of another sweetener would then be com-
> pared with that of sucrose, at doses also includ-
> ing maximum conditions of exposure under intended
> human use.

While not entirely new (relative and qualitative risk
comparisons among energy technologies abound[9]) Gori's

approach appears particularly well-suited to cases such as cancer hazards where establishment of causality is far from imminent. In addition, Gori's analysis stands in interesting contrast to the approach of Deisler (Chapter 15) who argues that, flawed as they are, animal experiments can be interpreted in terms of human effects and thus used in setting quantitative exposure standards for chemicals.

Chapters 9-11, then, contain a number of conceptual approaches for grouping and classifying hazards, and in this way they offer interesting possibilities for improving hazard management. Specific methodologies for management have, however, neither been spelled out nor tested by any of the authors. What is offered is therefore only the surface of a larger problem, the solution of which will require at the very least that the taxonomies proposed be both understood and accepted by people--including ordinary citizens--who must make decisions about hazard control.

References and Notes

1. "Water Quality Criteria," *Federal Register,* 44 (1979), 15926-15981; Toxic Substances Strategy Committee, *Toxic Chemicals and Public Protection* (TSSC, Washington, 1979), p. 108 (hereafter cited as *Toxic Chemicals).*

2. "National Emissions Standards for Hazardous Air Pollutants: Policy and Procedures for Identifying, Assessing and Regulating Airborne Substances, *Federal Register,* 44 (1979), 58642-58670.

3. "Regulations for the Enforcement of the Federal Insecticide, Fungicide, and Rodenticide Act, Subpart A: Registration, Re-Registration and Classification Procedures," *Federal Register,* 40 (1975), 28242-28246; "Interim Procedures and Guidelines for Health Risk and Economic Impact Assessments of Suspected Carcinogens," *Federal Register,* 41 (1976), 21402-21404.

4. "Identification, Classification and Regulation of Potential Carcinogens," *Federal Register,* 45 (1980), 5001-5296.

5. *Toxic Chemicals,* p. 111.

6. U.S. Interagency Regulatory Liaison Group, *Hazardous Substances* (IRLG, Washington, 1978).

7. *Regulatory Reporter,* 1 (June 1979)[+].

8. *Toxic Chemicals,* p. 114.

9. See, for example: H. Inhaber, *Risk of Energy Production* (Report AECB-1119, rev. 3; Atomic Energy Control Board, Ottawa, 1979); U.S. Nuclear Regulatory Commission, *The Reactor Safety Study* (WASH-1400, NUREG-75/014; Nuclear Regulatory Commission, Washington, 1975); National Research Council, Committee on Nuclear and Alternative Energy Systems, *Energy in Transition, 1985-2010* (W.A. Freeman, San Francisco, 1980), pp. 48-61; W. Ramsay, *Unpaid Costs of Electrical Energy* (Johns Hopkins University Press for Resources for the Future, Baltimore, 1979).

Christoph Hohenemser, Roger E. Kasperson,
Robert W. Kates

9. Causal Structure: A Framework for Policy Formulation

Controlling hazard in our technological society has become a major industry. Despite the positive benefits of technology, an increasingly common view has it that the benefits of technology exact their toll in environmental degradation, anxiety, illness, injury, and premature death. Especially in the last 10 years the American public has experienced a relentless parade of threats arising in technology. Some of these threats occur on the familiar "macroscopic" scale of oil spills, gas explosions, dam breaks, and air crashes; others appear on the intrinsically "molecular" scale of pesticides, food additives, and drugs. A few cases, which tend to arouse maximum fear, combine macroscopic and molecular scale threats as in the near meltdown of the reactor at Three Mile Island, or recurrent accidents involving fire and the simultaneous release of toxic chemicals.

We define hazards as threats to humans and what they value, and we take risks to be measures of these threats. A review of the scientific literature shows that scientists prefer to express risks as conditional probabilities for experiencing harm and frequently seek to use these probabilities as bases for prescribing societal response. Lay people, in contrast, judge risk on more complex scales, including such qualities as catastrophic potential, voluntary/involuntary character, and the degree to which risks are familiar and unfamiliar (Slovic et al, Charter 10).

The degree of control that society achieves over particular hazards depends strongly on the character of institutions and laws, as well as perceptions of individuals. Consequently, a narrow definition of risk as a conditional

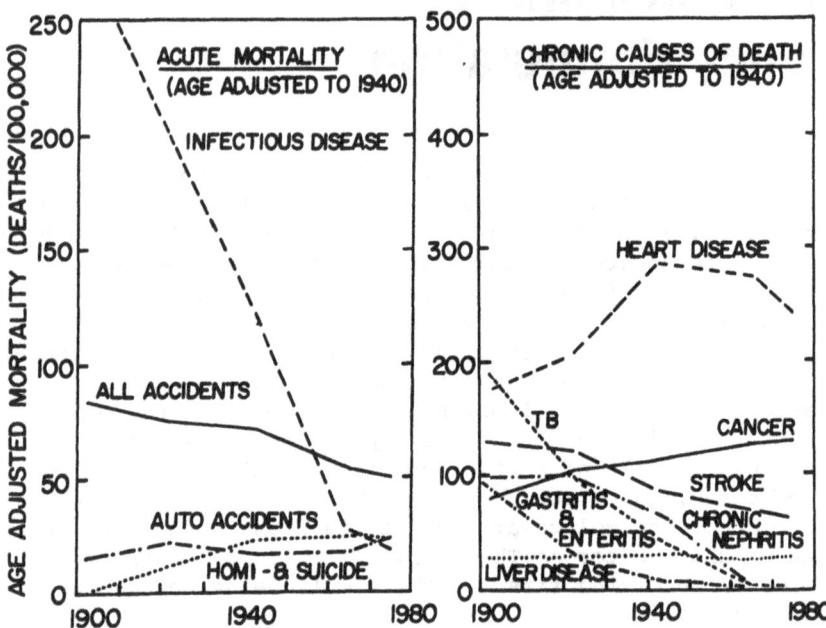

Figure 9.1. Historical variation of age-adjusted causes of death in the United States from 1900 to the present. Among acute causes, note the sharp decline of infectious disease and the rise in auto-accident mortality; among chronic causes of death, note the decline of most diseases except for cancer and cardiovascular disease.

probability is unlikely to predict individual or societal
response. Instead, one finds that societal effort expended
on risk control varies widely from case to case. Nevertheless, we believe that the generic problems of hazard control
are rooted in the structure and the physical dimensions of
hazards. As an introduction to this structure, we consider
in this paper: (1) the magnitude of the technological hazard
burden, as measured by hazard consequences; (2) the anatomy
of hazards, as described by a generally applicable "causal
model"; and (3) a taxonomy of technological hazard based on
this causal model. Taken together, these three approaches
lead to a first-order picture of technological hazard that
may be useful in further illuminating research as well as in
informing public policy.

The Burden of Technological Hazard

It is well to remember at the outset that technology is
associated with both decrease and increase of risk. Age-
adjusted mortality time trends (Figure 9.1) show that deaths
from infectious disease, all accidents, and most intestinal
diseases decreased during the last 80 years, whereas deaths
from automobile accidents, cancer, and heart disease
increased.

In approaching the risk of a technological society, it
is also important to realize that as technology has grown,
the last 80 years in America have witnessed a net increase
of life expectancy of 20-30 years. This gain may be attributed to elimination of infectious disease, and as a necessary
consequence, the "transfer" of mortality from infectious
disease to causes of death that occur later in life.

In estimating today's burden of technological hazard,
we have limited ourselves to mortality as hazard consequence,
because data on it are most readily available. To determine
the fraction of mortality associated with technology, we
recognized at the outset three possible approaches, each
widely described in the literature:

(1) Mechanistic models that provide a direct link
    between technological energy and materials
    releases and human harm, as in the saturation
    of blood with carboxyhemoglobin in the case of
    carbon monoxide exposure.

(2) Correlative exposure/consequence data relating
    specific energy and materials releases with
    specific expression of human harm, as in the

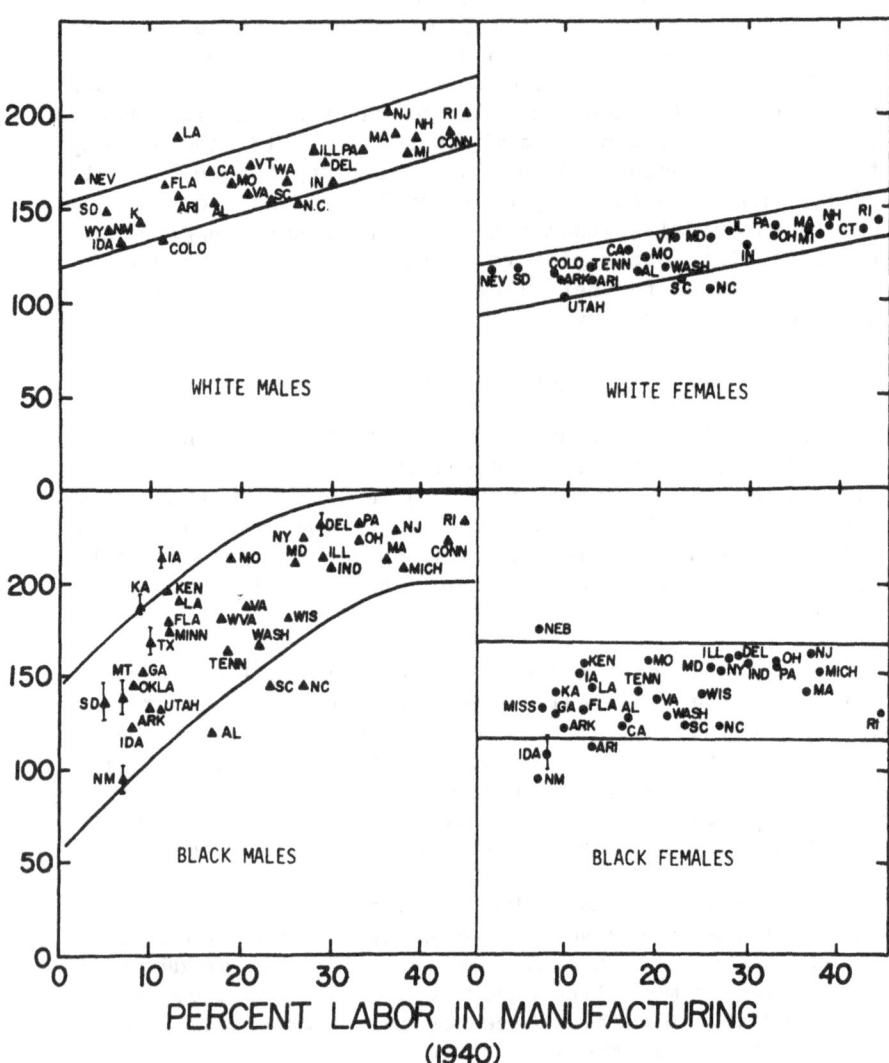

Figure 9.2. Correlation between average 1950-1969 age-adjusted cancer mortality and percent labor in manufacturing in 1940 for states within the United States. Note the pattern of increasing mortality with increasing industrial exposure except for black females.

link between exposure to ionizing radiation
and incidence of leukemia in humans.

(3) Definition of the fraction of exogenously "caused"
chronic disease through a comparison of highest
and lowest observed rates, as is done when
80-90% of cancer is assigned to "environmental
causes."

Upon reflection we regarded each of these approaches
inadequate. Approaches (1) and (2) require summing over a
wide range of kinds of exposure, only a small fraction of
which can be said to be defined through adequate data. It
is thus a foregone conclusion that such summing cannot induce
a complete answer. In contrast, approach (3) gives only an
upper limit for the desired quantity and is likely to be a
serious over-estimate.

To solve the problem at least in a rudimentary way, we
took a fourth approach.[1] Starting with rates of "standard
causes of death" as published by the World Health Organization[2] and the U. S. Department of Health, Education, and
Welfare,[3] we correlated mortality rates with technological
indicators such as percent employment in industry, per capita
energy consumption, and per capita gross national product
(GNP). In this way we learned that within the U. S. and
internationally, age-adjusted cancer mortality rates are
strongly related to level of technology whereas similar data
for heart disease, stroke, respiratory and liver disease are
not. Figure 9.2 illustrates our results for cancer in blacks
and whites, both male and female.

We thus conclude that the mortality risk associated with
technology may, at least in the case of cancer, be quite
substantial, the overall beneficial time trends of Figure
9.1 notwithstanding. We drew three specific conclusions from
our analysis of mortality:[1]

(1) Hazards of technology have in the industrial
nations replaced ancient natural hazards of
floods, pestilence, and disease (Table 9.1).

(2) In the United States, the death toll associated
with technological hazards is 20-30% of all male
mortality and 10-20% of all female mortality
(Table 9.2).

(3) In the United States, the total value of medical
costs and lost productivity was $50-75 billion
in 1974, with about half of the total connected

Table 9.1. Comparative hazard sources in U.S. and developing
countries

	PRINCIPAL CAUSAL AGENT[a]			
	NATURAL[b]		TECHNOLOGICAL[c]	
	Social cost[d] (% of GNP)	Mortality (% of total)	Social cost[d] (% of GNP)	Mortality (% of total)
United States	2–4	3–5	5–15	15–25
Developing countries	15–40[e]	10–25	n.a.[f]	n.a.[f]

a. Nature and technology are both implicated in most hazards.
   The division that is made here is made by the principal
   causal agent, which, particularly for natural hazards, can
   usually be identified unambiguously.
b. Consists of geophysical events (floods, drought, tropical
   cyclones, earthquakes, and soil erosion); organisms that
   attack crops, forests, livestock; and bacteria and viruses
   which infect humans. In the U.S. the social cost of each
   of these sources is roughly equal.
c. Based on a broad definition of technological causation,
   as discussed in the text.
d. Social costs include property damage, losses of productivity from illness or death, and the costs of control
   adjustments for preventing damage, mitigating consequences,
   or sharing losses.
e. Excludes estimates of productivity loss by illness, dis-
   ablement, or death.
f. No systematic study of technological hazards in developing
   countries is known to us, but we expect them to approach
   or exceed U.S. levels in heavily urbanized areas.

Table 9.2. Estimates of technologically involved deaths in the United States

| Cause of death | Percent male | Percent female | Annual deaths in thousands | |
			male	female
ACUTE MORTALITY				
Infectious disease	0	0	0	0
Deaths in infancy	5	5	1	1
Transportation accidents	90	90	39	15
Other accidents	70	50	28	11
Violence	30	30	10	3
Other acute deaths	70	50	8	5
CHRONIC MORTALITY				
Cardiovascular disease	0-40	0-40	0-217	0-132
Cancer	40	25	82	35
Chronic liver disease	0	0	0	0
Chronic respiratory disease	0-20	0-5	0-5	0
Other chronic disease	25	25	19	15
ALL MORTALITY	17-30	11-21	182-318	85-167

For a description of the methods used in obtaining the above results, see Harriss, et al.[1]

Table 9.3. Technological hazard control costs and damage for fiscal year 1979*

Description	Cost (Billions of $)
Federal control costs	$ 22-35
State and local control costs	11-17
All public (federal and state) control costs	32-52
Private control costs	67-80
Total government and private control	99-133
Public and private damage costs	80-150
Total control and damage costs	179-283

* For a description of the methods used in obtaining the results shown here, see the work of J. Tuller.[4]

with accidents and violence and the other half
with chronic disease.

Although individuals enjoy some choice and control over
the hazards they face, their responses are substantially
constrained. Many risks are involuntary, involve difficult
tradeoffs between benefit and risk, and are characterized
by insufficient individual knowledge to permit sound individual decision making. To compensate society attempts to fill
the gap. Hence, government and industry have increasingly
usurped the individual's role in managing risk. Our study
of this societal effort shows it to be extensive. Federal
spending on hazard management and control was estimated to
be $22-35 billion per year in 1979, whereas public and pri-
vate damage costs totalled $80-150 billion. A detailed
breakdown appears in Table 9.3.

The Causal Anatomy of Technological Hazard

If we are to understand hazards beyond the mortality
they produce, it is essential to inquire into their structure.
Accordingly, we present here a model of technological hazards
that we believe is simple enough to serve as a framework for
policy analysis and sufficiently detailed to provide an
adequate representation of reality. As will be seen, our
model characterizes the chain of causation that leads to
human harm, while sidelining for the sake of simplicity any
explicit reference to the benefits of technology.

Our way of thinking of hazard causation grew out of a
number of years of work on natural hazards.[5] Put simply,
our approach divides hazards into two components, events and
consequences. In this characterization hazard events represent the potential for harm and hazard consequences the
realization of harm. Consequences are measured in a variety
of ways, including death, injury, economic and social loss.
The separation of hazard into two components suggests three
ways of coping or managing: (1) prevention of hazard events;
(2) prevention of hazard consequences once events have
occurred; and (3) mitigation of consequences once these have
occurred.

A more versatile form of the model divides events into
two parts, initiating events and outcomes. A further elaboration recognizes pathways connecting stages of hazard
development. An illustration of the model in this form
appears in Figure 9.3. Note the arrow of time, indicating
the sequence of hazard development. Pathways connecting
stages represent possible points for exercising hazard control, as indicated. Note further that for the case illus-

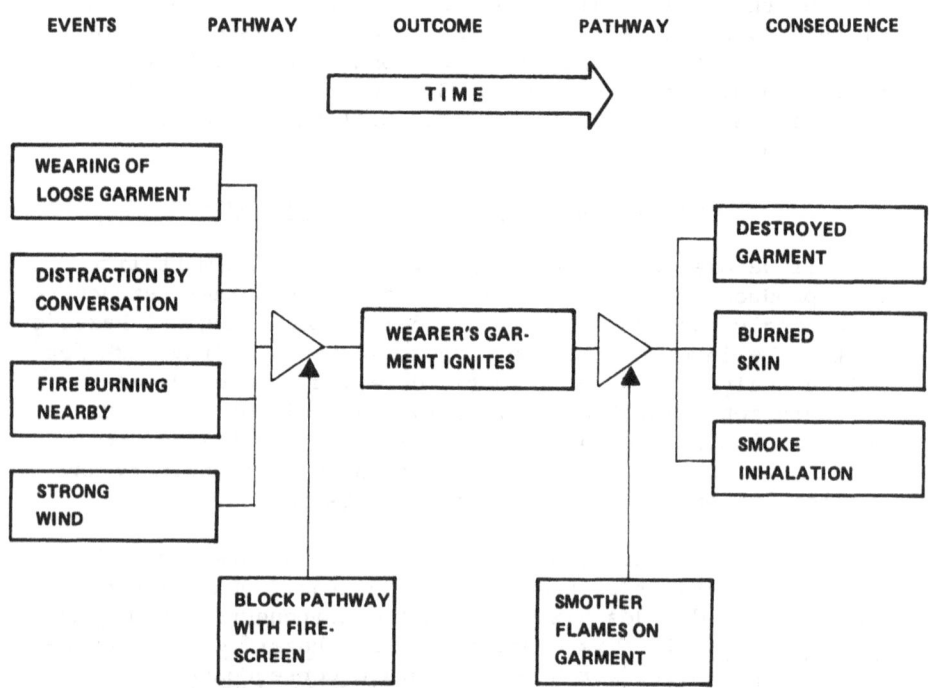

Figure 9.3. Three-stage model of hazard causation. The case illustrated involves a garment that accidentally ignites when the wearer stands too near an open fire.

trated, several parallel events are required more or less
simultaneously in order for initiating events to evolve to
a subsequent outcome.

The model may be expanded further. For example, just as
outcomes follow initiating events, higher order consequences
may follow first order consequences, all appropriately linked
by pathways. From a practical point of view, additional stages
are introduced because they define additional meaningful management opportunities. Figure 9.4 provides two examples of
expansion. In the first, several orders of consequences are
indicated to show how a burn may lead to eventual death; in
the second, several orders of outcomes are sketched to show
how a corroded brake lining can lead to an auto crash.

Finally, to take the model to the origin of hazards, we
add three stages upstream of initiating events. These we
term choice of technology, human wants, and human needs. As
illustrated in Figure 9.5 diagramming this full scope of
hazard causality is particularly important in cases, such as
pesticide use, for which downstream options for hazard management are poorly understood or unavailable.

For any given technology, full expansion of the model
may be a major undertaking that can lead to baroque structures. Such structures may be fun to build, but they will
have little use for decision-makers. We think, therefore,
that expansion should be restricted to 10 or fewer stages
if policy analysis is intended. In that form, the model can
be of use to decision-makers in a number of ways. We consider these next.

Expanded understanding. The most immediate advantage
of filling in the logical structure of a hazard is the fuller
understanding that is gained. Thus, the model forces us to
look at all logically possible management options. For
example, the diagramming of pesticide use (Figure 9.5) led
to our identification of a potential intervention--"to prevent cancer after ingestion of contaminated fish"--
characterized as "method unknown." Indeed, this intervention
remains, to our knowledge, impossible, but it is nonetheless
worthy of research.

Mapping regulatory effort. The model may be used to
map regulatory effort applied at each stage of control intervention. In most cases this immediately raises obvious
questions about distribution of effort. A level-of-effort
map requires convenient indicators, such as manpower, budget
allocations, or the number of regulatory standards issued.
To illustrate, we show in Figure 9.6 the distribution of

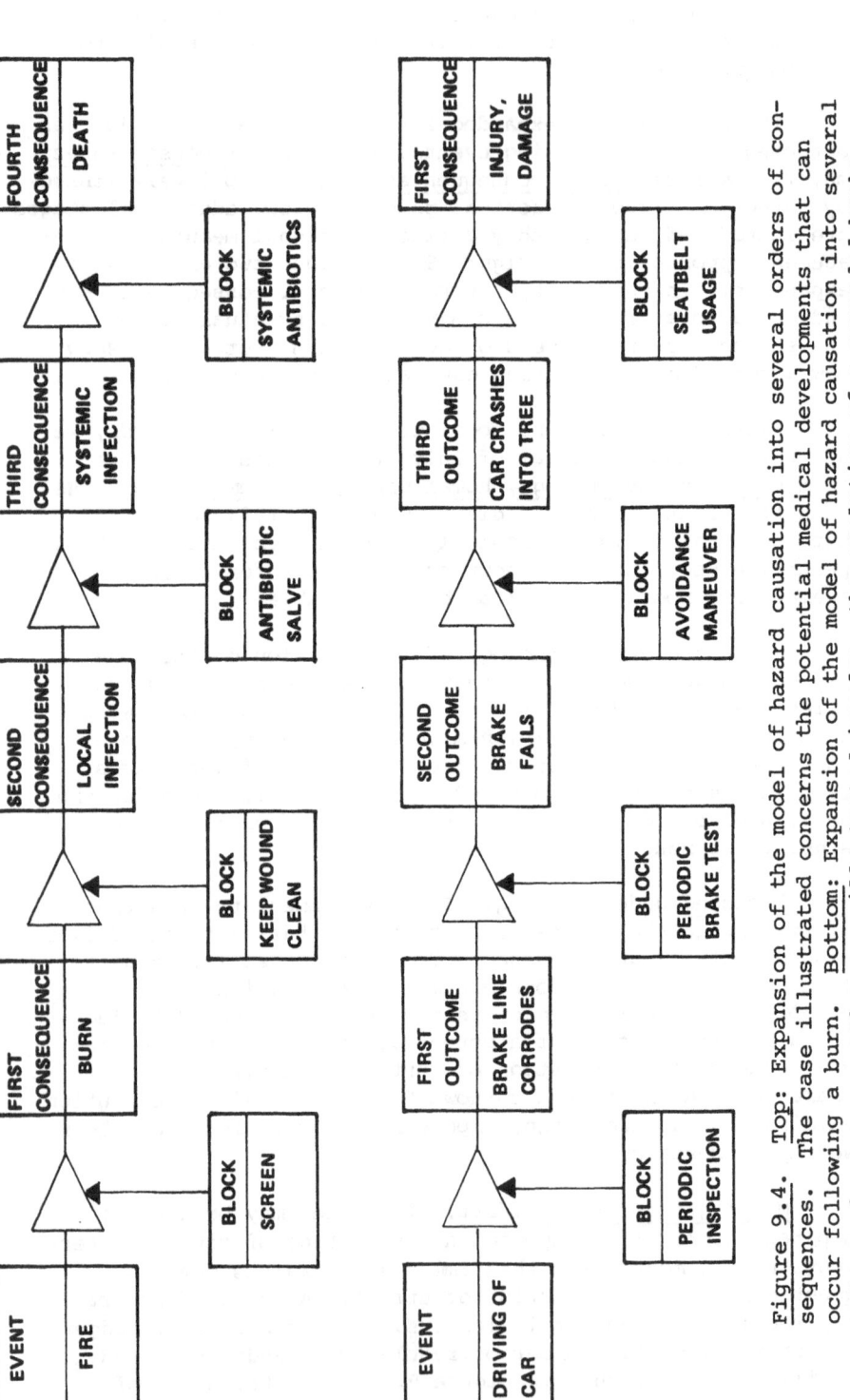

Figure 9.4. Top: Expansion of the model of hazard causation into several orders of consequences. The case illustrated concerns the potential medical developments that can
occur following a burn. Bottom: Expansion of the model of hazard causation into several
orders of outcomes. The case illustrated involves the evolution of a corroded brake
lining into a car crash.

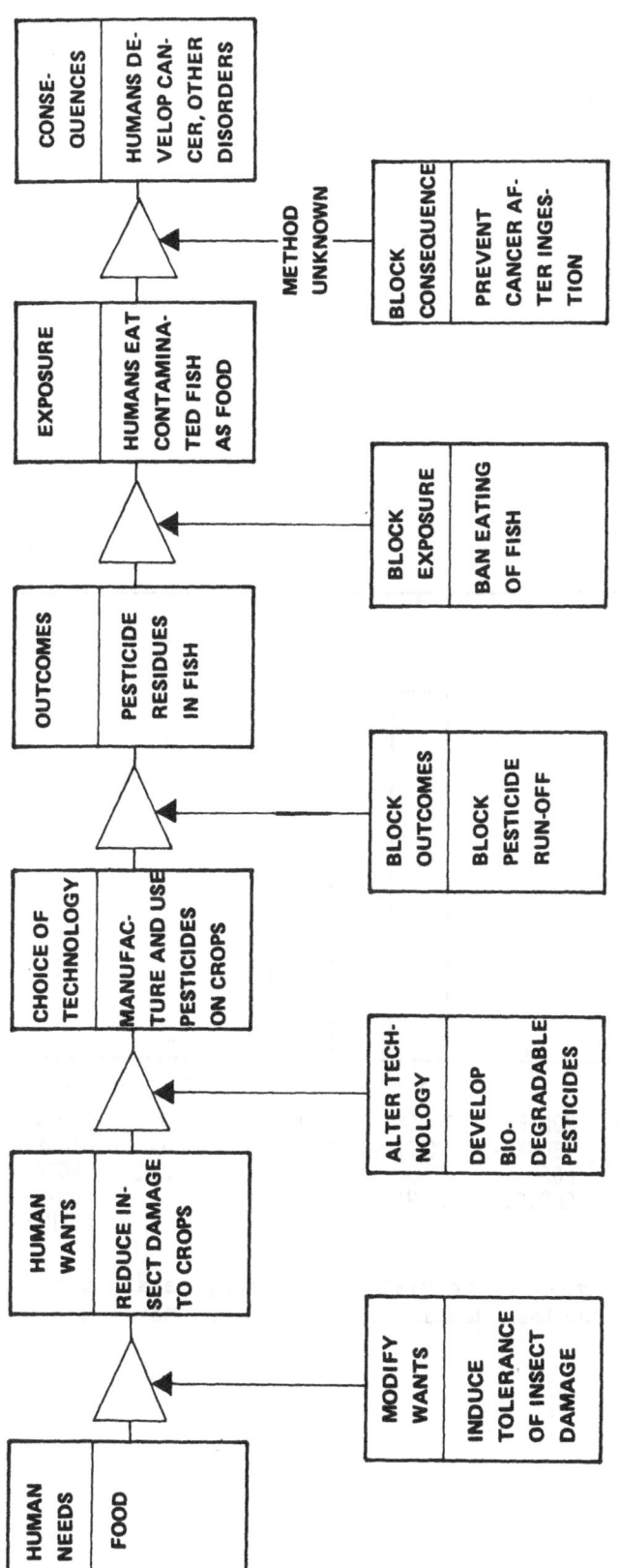

Figure 9.5. Expansion of the model of hazard causation into the full range of stages extending from human needs to consequences. The case illustrated involves the use of pesticides to suppress crop damage. It serves as a good example of the situation in which "downstream" management options involving events and consequences are not very promising or even possible, and "upstream" options involving human wants and choice of technology are most likely to succeed.

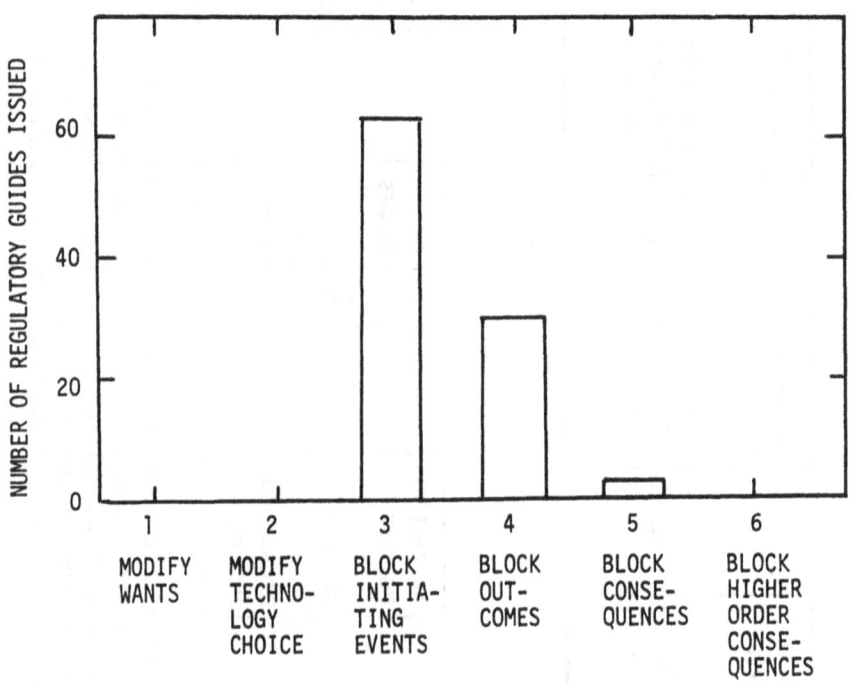

Figure 9.6. Number of regulatory guides by hazard stage issued by the Nuclear Regulatory Commission (Data through 1975).

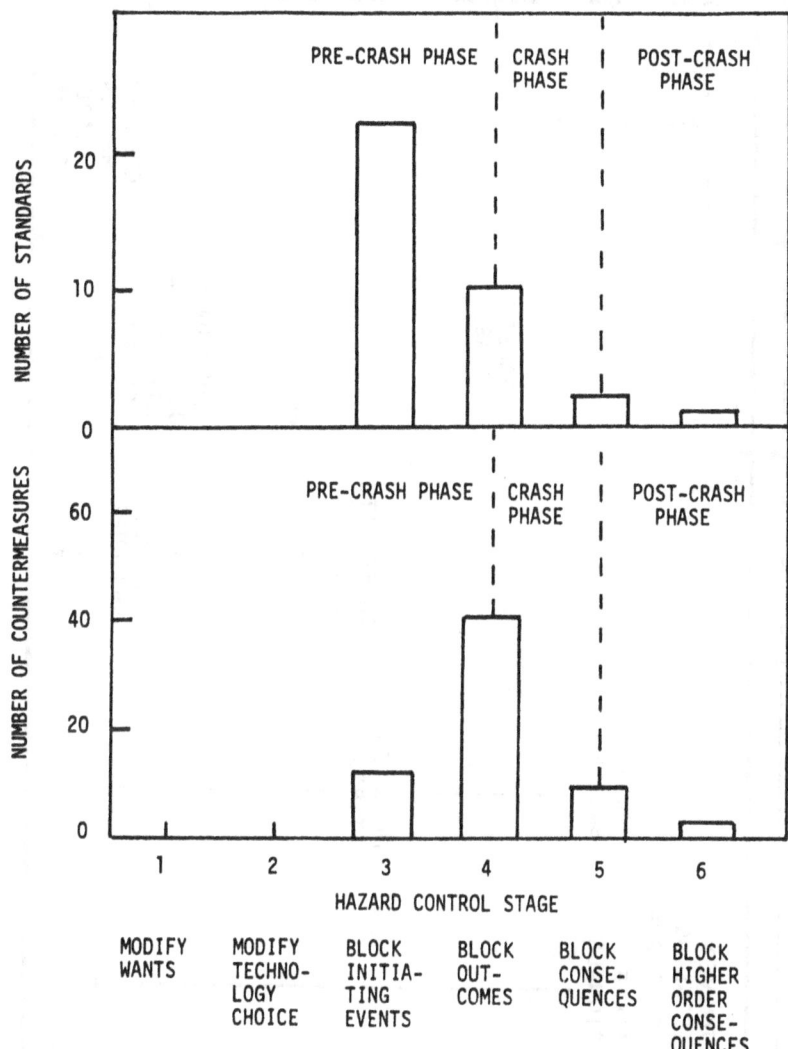

Figure 9.7. Top: Highway safety standards issued by the
Department of Transportation by hazard control stage.
Bottom: Highway safety "countermeasures" envisioned in
the 1976 highway safety report published by the Department
of Transportation.[6]

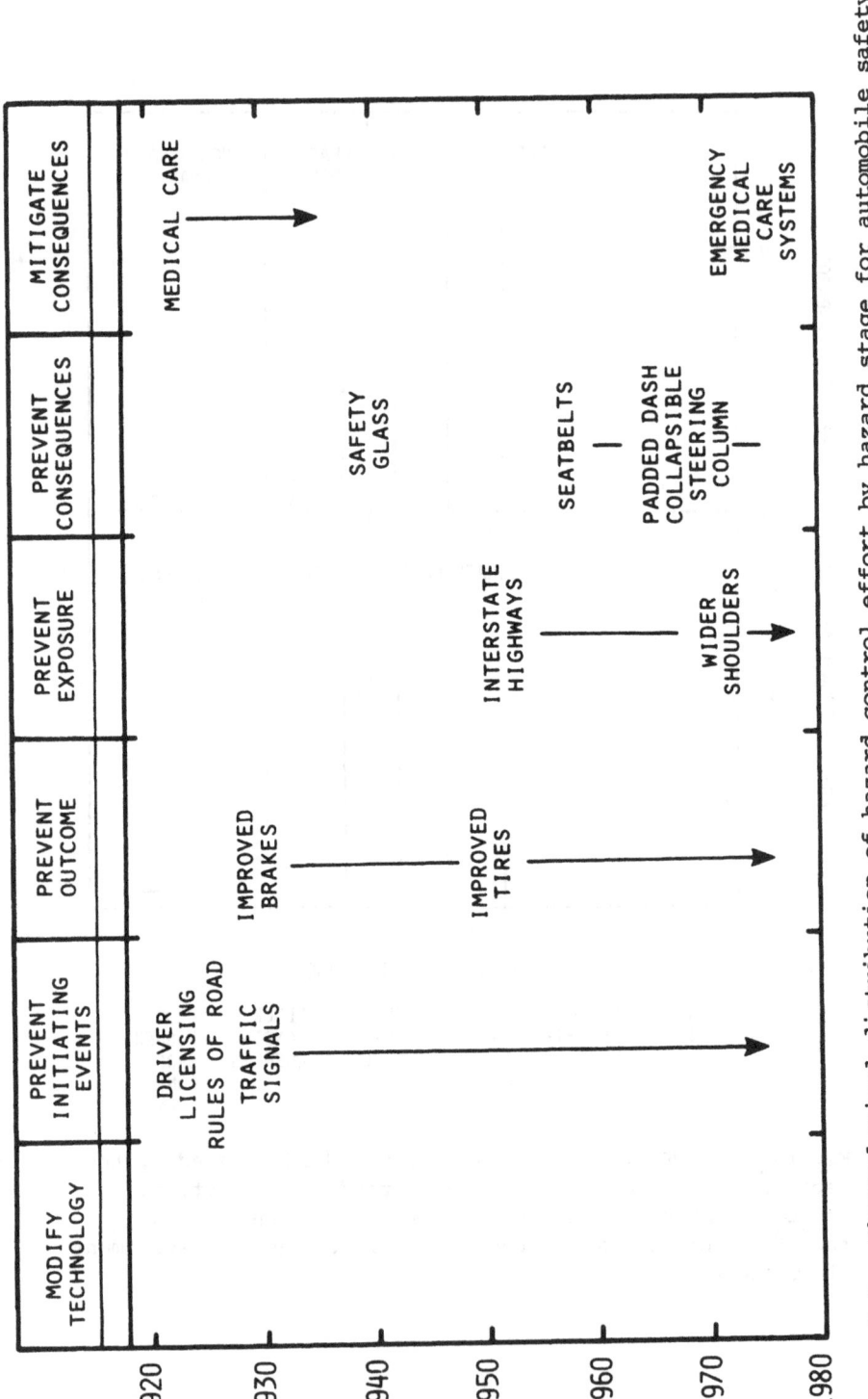

Figure 9.8. Chronological distribution of hazard control effort by hazard stage for automobile safety.

regulatory guides issued by the Nuclear Regulatory Commission through 1975 on the question of reactor safety. Of the 95 guides issued, 63 focussed on initiating events, 29 on outcomes, 3 on consequences, and none on consequence mitigation. A good example of consequence mitigation is evacuation to reduce radiation exposure. Since Three Mile Island, this has belatedly become a major priority, one which our model and the effort map in Figure 9.6 enabled us to recognize in 1976.

A similar level-of-effort map (Figure 9.7, top) categorizes highway safety standards issued by the U. S. Department of Transportation. It is readily apparent that 81% of the standards are classed as blocking initiating events, with little activity downstream. In contrast, a landmark highway safety report[6] analyzing 37 possible highway safety countermeasures places 40% of activity downstream from initiating events (Figure 9.7, bottom). Effort maps thus illuminate the difference between highway safety practice and highway safety theory and lead one to ask a number of questions. Is the distribution of effort appropriate for the physical nature of the hazard, the mandate and past history of the Department of Transportation, the perception of risk managers, or other factors? Is the distribution of effort optimal?

Classifying societal response. Societal response to hazard probably follows patterns in time that are structurally similar for similar hazards. To begin testing this hypothesis we have mapped control actions chronologically by hazard stage.

In the case of auto safety management, a schematic representation (Figure 9.8) shows that except for medical care administered to crash victims, the dominant early modes of management occurred far "upstream." The attempt to block injuries (first order consequences) once crashes have occurred is a rather recent development.

A diametrically opposite pattern emerges in the response of Japanese society to Minamata disease (Figure 9.9). Here control strategy begins downstream and in time moves steadily upstream, leading finally to the elimination of the technology, or what is the equivalent, its transfer to Thailand.[7]

Feedback analysis. In discussing the evolution of hazard from ultimate causes to final consequences, we have indicated control points without considering the details of the control processes. In many cases of hazard management, the simple flow of time from "upstream" to "downstream" is an inappropriate description. Instead, a hazard is first

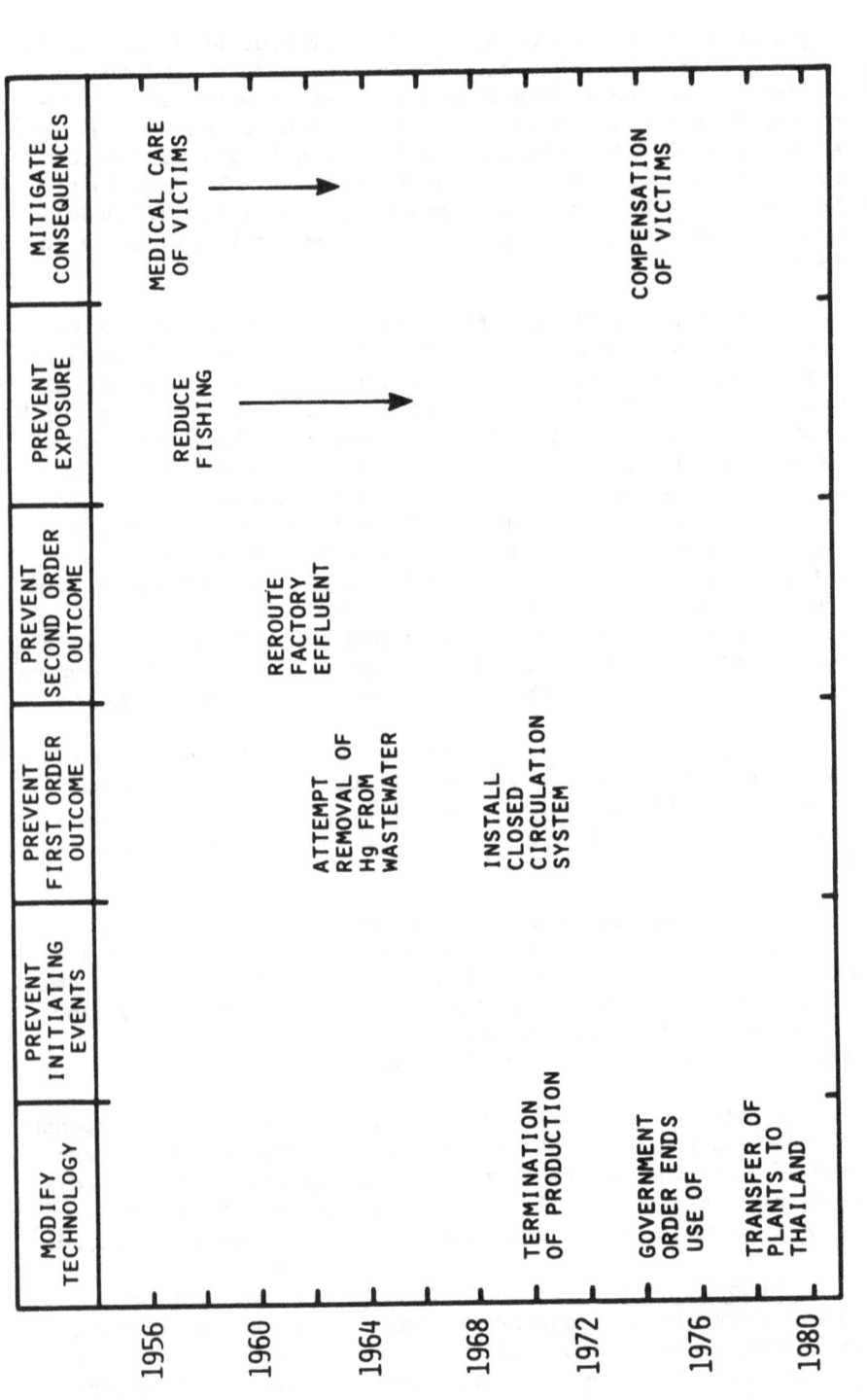

Figure 9.9. Chronological distribution of hazard control effort by hazard stage for Minamata disease.

recognized through an experienced outcome or consequence, and control action follows in time by inserting a block at appropriate upstream stages. In this sense, control intervention involves feedback, and information flows backward from downstream to upstream stages.

A great deal of feedback takes place in the way individuals strike a balance between hazard creation and reduction. Thus it is possible for effective feedback to occur through a "free market" mechanism, or through explicit regulatory actions.

Feedback may, in principle, be both positive or negative. For reducing hazard, we desire negative feedback; that is, we seek upstream control intervention that blocks or reduces consequences. Unfortunately, hazard management has in many cases produced unintended positive feedback, or processes through which upstream control interventions increase the level of consequences.

To illustrate both positive and negative feedback, we diagram in Figure 9.10 the case of flood damage control. The management sequence runs roughly as follows:

(a) Floods occur (consequences).

(b) Assessors prescribe flood protection in the form of dams and levees (initiation of feedback loop).

(c) Engineers build dams and levees (implementation of feedback).

(d) The dams prevent a number of floods (intended negative feedback achieved by blocking outcomes).

(e) Unanticipated by the assessors and engineers, individuals perceive increased safety in the newly protected floodplains and settle there (initiation of positive feedback).

(f) Eventual flood damage is greater than before building of dams because infrequent, but nevertheless possible, overtopping leads to catastrophic flood damage in built-up flood plains (positive feedback overwhelms negative feedback).

A more recent and equally important case of unintended positive feedback has emerged from the Kemeny Commission report on the accident at Three Mile Island.[8] A diagram

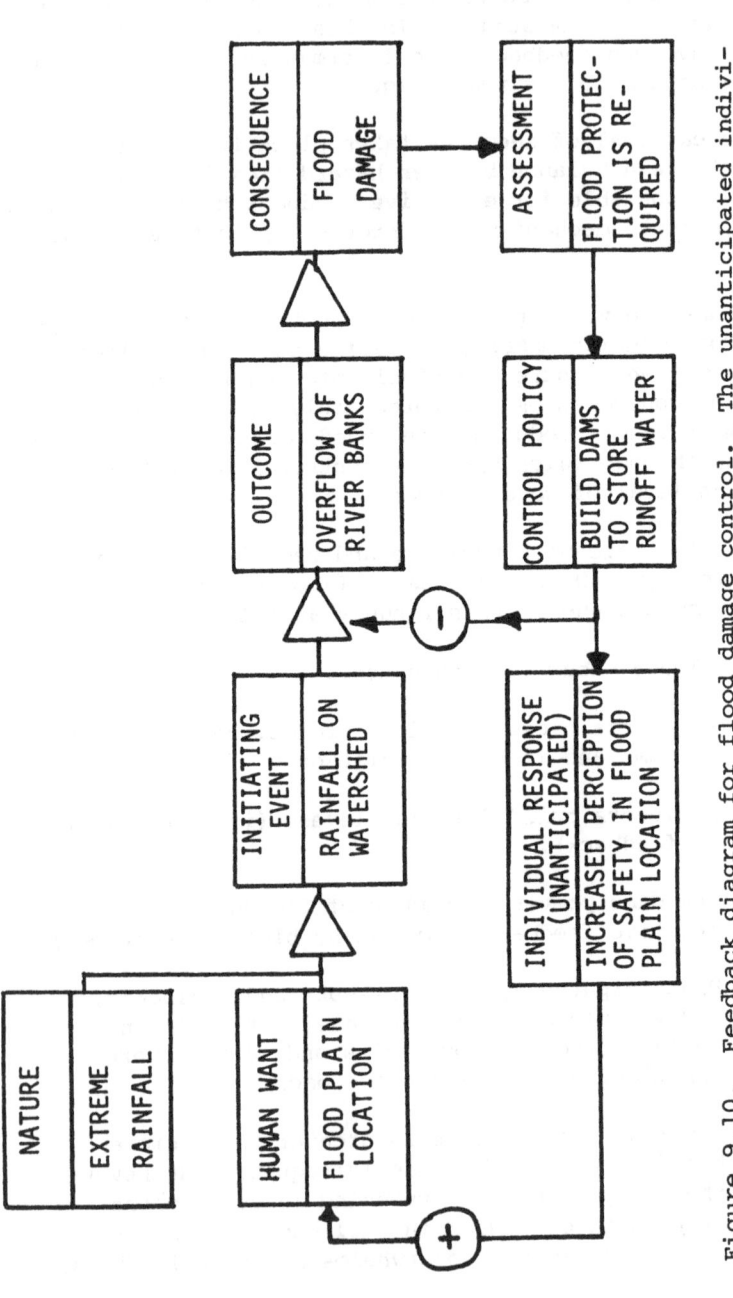

Figure 9.10. Feedback diagram for flood damage control. The unanticipated individual response of increased perceived safety in floodplain locations acts as a positive feedback and defeats the control policy.

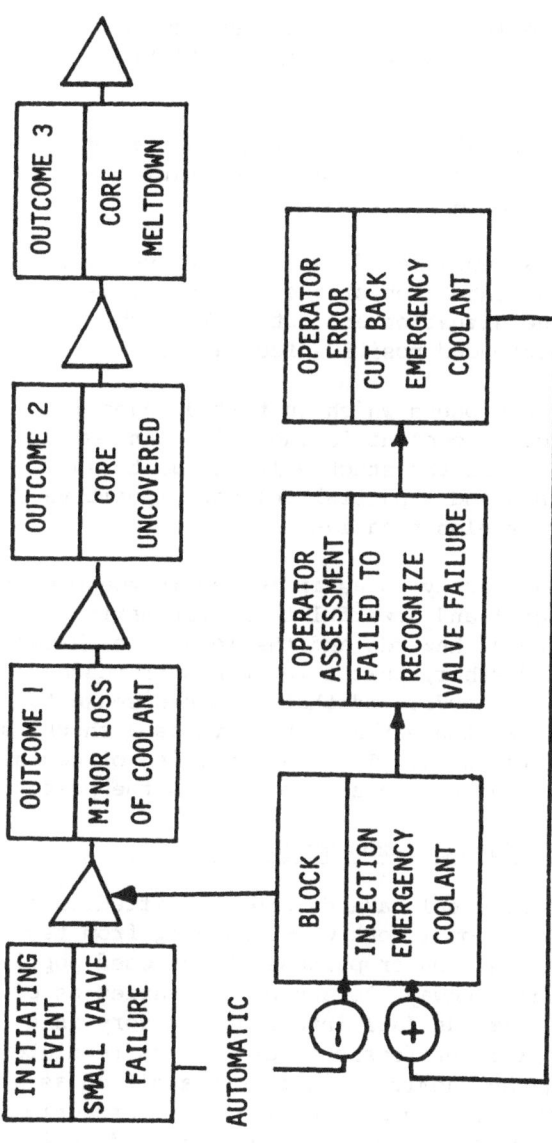

Figure 9.11. Partial feedback diagram for the accident at Three Mile Island. Negative feedback in the form of emergency core coolant was automatically triggered by the failure of a small valve in the primary cooling system. This was followed by an improper operator assessment and led to cutback of emergency coolant through operator action. The result was that the automatically triggered feedback was cancelled, and the initiating event progressed unchecked to a partial core meltdown.

appears in Figure 9.11. The corresponding sequence of events may be described as follows:

(a) A minor valve in the primary coolant system accidentally remains stuck open (initiating event).

(b) Through an automatic process, this initiates injection of emergency coolant (automatic block: negative feedback).

(c) The operators, unaware of the stuck valve, but observing the injection of the coolant, misinterpret the situation and cut back on the coolant (unintended positive feedback).

(d) The automatic process which initiated injection of the emergency coolant is cancelled, and over a period of hours the stuck valve leads to an uncovered core and a partial meltdown (positive feedback overwhelms negative feedback).

We regard feedback analyses as indicated in these examples as extremely useful and powerful tools for making explicit the structure of hazards and the methods of hazard control. We expect, further, that a few kinds of feedback diagrams apply to all hazards, and that it makes sense to construct a feedback catalogue that can serve as a checklist and guide for hazard managers. The challenge is, of course, to predict the future and not merely to explain the past.

Toward a Causal Taxonomy of Hazard

Currently, technological hazards are classified in a variety of apparently inconsistent ways, ranging from technological source, to function or purpose, to exposed population, to environmental pathway, to any or all varieties of consequences. Which hazards fall into what category is a function of historical or professional as well as regulatory organization. Any given chemical might be classified as a toxic substance, a threat to worker health, or a prescription drug. Integration of widely varying hazard classifications has been slow.

For the purpose of optimum management, it may be helpful to classify hazards in a more logical manner. Such a classification should allow use of similar managerial tools for all hazards of a given class. A successful classification will also facilitate more effective comparative assessment than is now possible. One should be able to answer such questions

as: what do saccharin, skateboards, and the collapse of the
Grand Teton Dam have in common? How do they differ?

Dimensions of causal structure. Recently we have used
the causal model described in Section 2 to construct a taxonomy of technological hazards. On the basis of information
derived from the scientific literature, we have coded 93
technological hazards on 16 dimensions. Each stage of the
causal structure of hazard is characterized by one or more
dimensions, as indicated in Figure 9.12. The scales used to
define each dimension are given in Table 9.4. Wherever
possible, we used numerical scales of logarithmic character:
scale increments of unity were defined to correspond to
multiplicative factors of 10 or 100. In this sense our
scales are similar to other sociophysical scales in which
physical events of human interest cover many orders of magnitude in physical "intensity." In the field of hazard
analysis, a well-known example of such a scale is the Richter
scale for earthquake intensity, on which increments of one
correspond to a factor of 10 in energy release. Beyond the
wide range of the physical dimensions underlying causal
structure, we chose logarithmic scales for two additional
and independent reasons: (1) given the paucity of information
for many hazards, it is unrealistic in general to differen-
tiate among hazards to any greater accuracy than a factor of
10; (2) in many cases, individual events in a given hazard
structure involve a range of values that may cover a factor
of 10 or more. Whether such crude scaling could capture interesting differences between hazards was a question we sought
to answer by trying the method. At the outset we had no a
priori insight that "factor-of-10 scaling" would be successful.

Hazard codification. Our initial selection of 66
hazards drew upon an existing library of case studies at
Clark University's Center for Technology, Environment, and
Development (CENTED), the caselist employed by Slovic and
his collaborators (Chapter 10), and informal discussion
within our group. Our early choices, then, were not supported by a systematic selection method, but they did include
a large fraction of the cases that had received public attention. After scoring the initial set of 66, we plotted their
distribution on each of the scales and noted the extent of
population imbalances. In selecting further hazards to round
out our sample, we made a special effort to correct such
imbalances. Our final sample of 93 is therefore reasonably
well distributed on most scales, though it can but reflect
the fact that there are few hazards in the extreme regions
of most scales, and many at the low end. Though our interest
is technological risk and hazard, we included in our sample
several "marker" cases related to smoking and alcohol use.

Table 9.4. Elementary variables

Symbol	Name	Variable Range	Variable Scale
G_1	Knowledge	poor to excellent	1 - 9
T_1*	Intentionality	"not intended to harm organisms" to "intended to harm humans."	3, 6, 8
T_2	Market value	$\$10^1 - 10^{12}$	1 - 7
T_3	Substitutability: broad	"no known substitutes" to "two or more"	1, 3, 5, 7, 9
T_4	Substitutability: specialized	"No known substitutes" to "two or more"	1, 3, 5, 7, 9
0_2*	Spatial extent	$1 - 10^{14}\ m^2$	1 - 9
0_3*	Concentration	energy: acceleration from 0 - 80 g.	1 - 9
		materials: concentration relative to background from 0 - 10,000+	

Table 9.4 continued

Symbol	Name	Variable Range	Variable Scale
0_4*	Persistence	1 min. - 20+ yrs.	1 - 9
0_5*	Frequency of recurrence	1 min. - 20+ yrs.	1 - 9
c_1*	Delay of conseq.	1 min. - 20+ yrs	1 - 9
c_2*	Maximum potential killed (U.S.)	$0 - 10^8 +$	1 - 9
c_3*	Population of risk (U.S.)	$0 - 10^8 +$	1 - 9
c_4*	Annual mortality (U.S.)	$0 - 10^8 +$	1 - 9
c_5*	Transgenerational	"no effect" to "several genera-tions affected"	3, 6, 9
c_6*	Non-human species mortality-max. potential	"none" to "signifi-cant mortality" to "species extinction"	3, 6, 9
c_7*	Non-human species mortality-experi-enced	"none" to "signifi-cant mortality" to "species extinction"	3, 6, 9

* Variables used in factor analysis

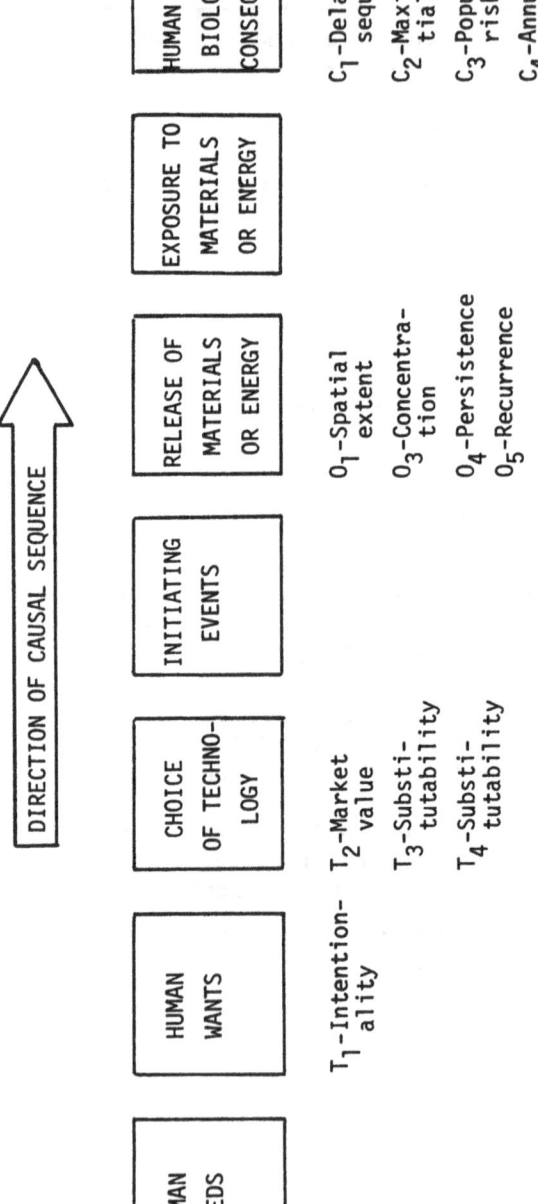

Figure 9.12. Causal model of hazard, with approximate location of variables. T-variables describe human wants and choice of technology. O-variables describe outcomes or releases. C-variables describe exposure-consequence relations and consequences. G_1 describes the state of knowledge of the whole chain.

Most hazards were scored by two or more individuals. Many cases were discussed in order to clarify the meaning of the available literature. After all scoring was complete, one individual made a series of checks for inconsistent scoring and thereby altered 20% of the scores by one scale point and a handful by as many as 2 or 3 scale points.

Composite Dimensions. Through factor analysis we have extracted from the 12 causal structure variables that describe hazard anatomy downstream from "choice of technology" five composite, orthogonal dimensions. The composite dimensions (factors) "explain" 82% of the variance of the sample. To good approximation this means that the causal structure of each of 93 hazards, and any others to be scored in the future, may be described by values of just five composite variables. Table 9.5 summarizes the relation of the composite variables to the original 12 variables.

The principal virtue of the analysis illustrated in Table 9.5 is that it simplifies thinking about hazards in a systematic way. The five composite variables are simpler to comprehend than the original 12. At the same time they constitute a considerable extension of the concept of technological risk frequently used by analysts. Only one of the five qualities--"annual mortality"--is normally included in quantitative discussions of risk; the other four--"intentionality," "persistence-delay," "catastrophic," and "global threat"--are dimensions that are normally omitted, or left as incidental remarks about risk.

A Taxonomy of Extremes. The factor analysis by itself is not a taxonomy of hazards, but it does offer several possible ways of constructing one. One such taxonomy may be derived from the factor analysis by identifying extreme scorers on each of the five composite scales as five hazard classes; multiple extremes as a sixth; and all others as a seventh and eighth class. The resulting grouping of hazards, illustrated in Table 9.6, has the effect of emphasizing the extremities of hazard space, on the assumption that it is these cases that merit special societal attention. In this sense, the present taxonomy defines in a systematic manner a partial list of society's "worry beads." Assuming that our methodology is found workable, this should remove some of the ad hoc character from much of the discussion of hazard identification, evaluation, and prioritization.

At present, society's struggle to deal with extreme scoring hazards is incomplete, though a well-identifiable policy exists for three classes. Thus, society prescribes prior approval, demonstrated safety, and restricted access

Table 9.5. Factor structure for combined energy and materials hazards: 93 cases

FACTOR			CAUSAL DIMENSIONS	
No.	Name	Variance explained	Name	factor loading
1.	Intentional	.32	C_7 non-human species mortality (experienced)	.87
			C_6 non-human species mortality (potential)	.79
			T_1 intentional design of technology	.81
2.	Persistence/ delay	.19	O_4 persistence of release	.81
			C_1 delay of consequences	.85
			C_5 transgenerational effects	.84
3.	Catastrophic	.11	O_5 rarity of occurence	.91
			C_2 maximum potential killed	.89
4.	Annual mortality	.10	C_4 annual mortality	.85
5.	Global threat	.09	C_3 population at risk	.73
			Concentration above background	.73

Table 9.6. Proposed taxonomy

	Classes	Examples
1.	Multiple Extremes	Nuclear war, Recombinant DNA, Deforestation
2.	Intentional Biocides	Pesticides, Nerve gas, Antibiotics
3.	Persistent Teratogens	Uranium mining, Radioactive Waste, Mercury
4.	Rare Catastrophes	Recombinant DNA, LNG, Satellites
5.	Common Killers	Automobiles, Handguns, Medical x-rays
6.	Diffuse Global Threats	Fossil fuel (CO_2), SST (NO_x), Coal Burning
7.	Macro Materials, Energy	Skateboards, SST (noise), Underwater Construction
8.	Micro Materials, Energy	Saccharin, Laetrile, Microwave Appliances

and use for intentional biocides; prior approval, event prevention and engineered safety for major catastrophic hazards;
and educational efforts, behavior modification, technical
fixes, and diffuse responsibility for common killers.
Society is seeking a policy for persistent, delayed tera-
togens--through debates on cancer principles, burden of
proof, validity of animal experiments, and liability. And
society is still assessing the risks for diffuse global
hazards such as atmospheric CO_2 buildup and acid rain.

From its structure, it is clear that our codification
of technological hazards via quantitative measures of causal
structure will permit a number of alternative interpretations.
The present, brief report on our hazard classification
efforts must therefore be regarded as a preliminary view of
a still developing story. Current and future work is concerned with validation of scoring procedures; with more
suitable classification of routine, non-extreme hazards; with
the relation between hazard dimensions and benefits; and with
a detailed comparison of our codification to perception.

Summary and Prognosis

We have estimated the magnitude of the technological
hazard problem and shown it to be substantial; described a
model of technological hazard that makes explicit the causal
structure; suggested several ways in which the model can be
useful in clarifying issues of hazard control; and described
how a taxonomy of hazard may be constructed.

The remaining question is: can our model of hazard, its
application to mapping effort and classifying hazards, improve hazard control? Our hope is that the answer is yes.
But at this stage proof of this is not yet in, nor will it
be until the causal model and the applications we have
suggested are adopted by people who must make decisions
about hazards in their roles as citizens, managers of private
industry, and government regulators.

References and Notes

1. R. C. Harriss, C. Hohenemser, and R. W. Kates, "Our Haz-
    ardous Environment," *Environment,* 20, no. 7 (September,
    1978), 6-13, 38-41.

2. *World Health Statistics Annual* (World Health Organiza-
    tion, Geneva, 1976), vol. 1.

3. *Vital Statistics of the United States* (Dept. of Health,

Education, and Welfare, National Center for Health Statistics, Rockville, MD, 1976).

4. J. Tuller, "Technological Hazards: The Economic Burden," in *Technology as Hazard*, ed. R. W. Kates and C. Hohenemser (Oelgeschlager, Gunn, and Hain, Cambridge, MA, forthcoming).

5. I. Burton, R. W. Kates, and G. F. White, *The Environment as Hazard* (Oxford University Press, New York, 1978).

6. U. S., Dept. of Transportation, *The National Highway Safety Needs Report* (Washington, DOT, 1976).

7. S. Suckcharoen, P. Nuorteva, and E. Häsänen, "Alarming Signs of Mercury Pollution in a Freshwater Area of Thailand," *Ambio*, 7, no. 3 (1978), 113-116.

8. U.S., President's Commission on the Accident at Three Mile Island, *The Need for Change: The Legacy of TMI* (Government Printing Office, Washington, 1979).

Paul Slovic, Baruch Fischhoff,
Sarah Lichtenstein

10. Rating the Risks:
The Structure of Expert
and Lay Perceptions

People respond to the hazards they perceive. If their perceptions are faulty, efforts at public and environmental protection are likely to be misdirected. In order to improve hazard management, a risk assessment industry has developed over the last decade which combines the efforts of physical, biological, and social scientists in an attempt to identify hazards and measure the frequency and magnitude of their consequences.[1]

For some hazards extensive statistical data are readily available; for example, the frequency and severity of motor vehicle accidents are well documented. For other familiar activities, such as the use of alcohol and tobacco, the hazardous effects are less readily discernible and their assessment requires complex epidemiological and experimental studies. But in either case, the hard facts go only so far and then human judgment is needed to interpret the findings and determine their relevance for the future.

Other hazards, such as those associated with recombinant DNA research or nuclear power, are so new that risk assessment must be based on theoretical analyses such as fault trees (see Figure 10.1), rather than on direct experience. While sophisticated, these analyses, too, include a large component of human judgment. Someone, relying on educated intuition, must determine the structure of the problem, the consequences to be considered, and the importance of the various branches of the fault tree.

Once performed, the analyses must be communicated to the

Reprinted by permission from Environment, vol. 21 (April 1979).

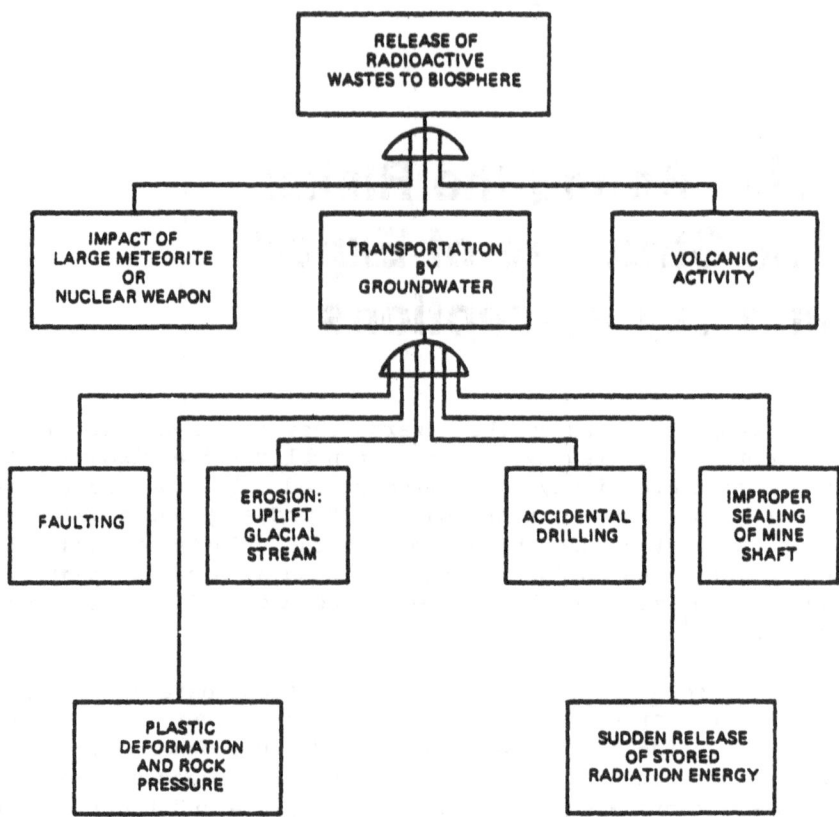

Figure 10.1.   Illustration of a fault tree.   Fault trees are used most often to characterize hazards for which direct experience is not available.   The tree shown here indicates the various ways in which radioactive material might accidentally be released from nuclear wastes buried within a salt deposit. To read this tree, start with the bottom row of possible initiating events, each of which can lead to the transportation of radioactivity by groundwater.   This transport can in turn release radioactivity to the biosphere.   As indicated by the second level of boxes, release of radioactivity can also be produced directly (without the help of groundwater) through the impact of a large meteorite, a nuclear weapon, or a volcanic eruption.   Fault trees may be used to map all relevant possibilities and to determine the probability of the final outcome.   To accomplish this latter goal, the probabilities of all component stages, as well as their logical connections, must be completely specified.   (Source: P. E. McGrath, "Radioactive Waste Management," Report EURFNR 1204, Karlsruhe, Germany, 1974.)   Reprinted by permission.

various people, including industrialists, environmentalists,
regulators, legislators, and voters, who are actually responsible for dealing with the hazards. If these people do not
see, understand, or believe these risk statistics, then distrust, conflict, and ineffective hazard management may result.
Ember's saga of the bungling at Love Canal (Chapter 7) is a
case in point.

Judgmental Biases

When lay people are asked to evaluate risks, they seldom
have statistical evidence on hand. In most cases they must
rely on inferences based on what they remember hearing or observing about the risk in question. Recent psychological research has identified a number of general inferential rules
that people seem to use in such situations.[2] These judgmental rules, known technically as <u>heuristics</u>, are employed
to reduce difficult mental tasks to simpler ones. Although
valid in some circumstances, in others they can lead to large
and persistent biases with serious implications for risk assessment.

Availability

One heuristic that has special relevance for risk perception is known as "availability".[3] People who use this
heuristic judge an event as likely or frequent if instances
of it are easy to imagine or recall. Frequently occurring
events are generally easier to imagine and recall than rare
events. Thus, availability is often an appropriate cue.
Availability, however, is also affected by numerous factors
unrelated to frequency of occurrence. For example, a recent
disaster or a vivid film such as <u>The China Syndrome</u> may
seriously distort risk judgments.

Availability-induced errors are illustrated by several
recent studies in which we asked college students and members of the League of Women Voters to judge the frequency of
various causes of death, such as smallpox, tornadoes, and
heart disease.[4] In one study, these people were told the
annual death toll (50,000) for motor vehicle accidents in the
United States; they were then asked to estimate the frequency
of forty other causes of death. In another study, participants were given two causes of death and asked to judge which
of the two is more frequent. Both studies showed people's
judgments to be moderately accurate in a global sense; that
is, people usually knew which were the most and least frequent lethal events. Within this global picture, however,
there was evidence that people made serious misjudgments,
many of which seemed to reflect availability bias.

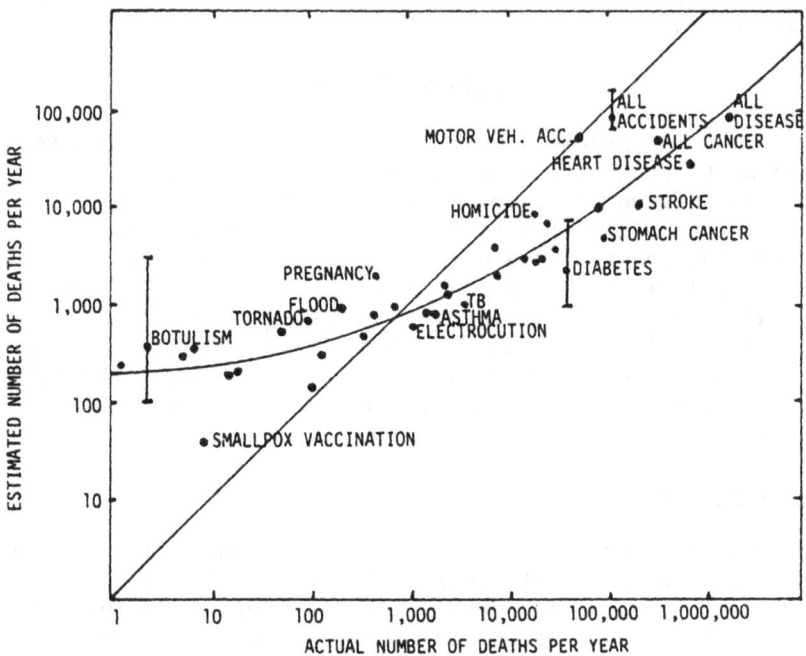

Figure 10.2. Relationship between judged frequency and the
actual number of deaths per year for 41 causes of death. If
judged and actual frequencies were equal, the data would fall
on the straight line. The points, and the curved line fitted
to them, represent the averaged responses of a large number
of lay people. While people were approximately accurate,
their judgments were systematically distorted. As described
in the text, both the compression of the scale and the scatter of the results indicate this. To give an idea of the degree of agreement among subjects, vertical bars are drawn to
depict the 25th and 75th percentile of individual judgment
for botulism, diabetes, and all accidents. Fifty percent of
all judgments fall between these limits. The range of responses for the other 37 causes of death was similar. From
Lichtenstein et al.[4] Copyright 1978 by the American Psychological Association. Reprinted by permission.

Figure 10.2 compares the judged number of deaths per year with the actual number according to public health statistics. If the frequency judgments were accurate, they would equal the actual death rates, and all data points would fall on the straight line making a 45-degree angle with the axes of the graph. In fact, the points are scattered about a curved line that sometimes lies above and sometimes below the line of accurate judgment. In general, rare causes of death were over-estimated and common causes of death were underestimated. As a result, while the actual death toll varied over a range of one million, average frequency judgments varied over a range of only a thousand.

In addition to this general bias, many important specific biases were evident. For example, accidents were judged to cause as many deaths as diseases, whereas diseases actually take about fifteen times as many lives. Homicides were incor-rectly judged to be more frequent than diabetes and stomach cancer. Homicides were also judged to be about as frequent as stroke, although the latter actually claims about eleven times as many lives. Frequencies of death from botulism, tor-nadoes, and pregnancy (including childbirth and abortion) were also greatly overestimated.

Table 10.1 lists the lethal events whose frequencies were most poorly judged in our studies. In keeping with availabi-lity considerations, overestimated items were dramatic and sensational whereas underestimated items tended to be unspec-tacular events which claim one victim at a time and are com-mon in nonfatal form.

In the public arena the availability heuristic may have many effects. For example, the biasing effects of memorabil-ity and imaginability may pose a barrier to open, objective discussions of risk. Consider an engineer's demonstrating the safety of subterranean nuclear waste disposal by pointing out the improbability of each branch of the fault tree in Figure 10.1. Rather than reassuring the audience, the presentation might lead individuals to feel that "I didn't realize there were so many things that could go wrong." The very discussion of any low-probability hazard may increase the judged proba-bility of that hazard regardless of what the evidence indi-cates.

In some situations, failure to appreciate the limits of "available" data may lull people into complacency. For ex-ample, we asked people to evaluate the completeness of a fault tree showing the problems that could cause a car not to start when the ignition key was turned.[5] Respondents' judgments of completeness were about the same when looking at

Table 10.1 Bias in Judged Frequency of Death

Most Overestimated	Most Underestimated
All accidents	Smallpox vaccination
Motor vehicle accidents	Diabetes
Pregnancy, childbirth, and abortion	Stomach cancer
Tornadoes	Lightning
Flood	Stroke
Botulism	Tuberculosis
All cancer	Asthma
Fire and flames	Emphysema
Venomous bite or sting	
Homicide	

the full tree as when looking at a tree in which half of the causes of starting failure were deleted. In keeping with the availability heuristic, what was out of sight was also out of mind.

Overconfidence

A particularly pernicious aspect of heuristics is that people are typically very confident about judgments based on them. For example, in a follow-up to the study on causes of death, participants were asked to indicate the odds that they were correct in their judgment about which of two lethal events was more frequent.[6] Odds of 100:1 or greater were given often (25 percent of the time). However, about one out of every eight answers associated with such extreme confidence was wrong (fewer than 1 in 100 would have been wrong if the odds had been appropriate). About 30 percent of the judges gave odds greater than 50:1 to the incorrect assertion that homicides are more frequent than suicides. The psychological basis for this unwarranted certainty seems to be people's insensitivity to the tenuousness of the assumptions upon which their judgments are based (in this case, the validity of the availability heuristic). Such overconfidence is dangerous. It indicates that we often do not realize how little we know and how much additional information we need about the various problems and risks we face.

Overconfidence manifests itself in other ways as well. A typical task in estimating failure rates or other uncertain quantities is to set upper and lower bounds so that there is a 98 percent chance that the true value lies between them. Experiments with diverse groups of people making many different kinds of judgments have shown that, rather than 2 percent of true values falling outside the 98 percent confidence bounds, 20 percent to 50 percent do so.[7] People think that they can estimate such values with much greater precision than is actually the case.

Unfortunately, experts seem as prone to overconfidence as lay people. When the fault tree study described above was repeated with a group of professional automobile mechanics, they, too, were insensitive to how much had been deleted from the tree. Hynes and Vanmarcke[8] asked seven "internationally known" geotechnical engineers to predict the height of an embankment that would cause a clay foundation to fail and to specify confidence bounds around this estimate that were wide enough to have a 50 percent chance of enclosing the true failure height. None of the bounds specified by these experts actually did enclose the true failure height. The multi-million dollar Reactor Safety Study (the "Rasmussen Report"),[9] in assessing the probability of a core meltdown in a nuclear reactor, used a procedure for setting confidence bounds that has been found in experiments to produce a high degree of overconfidence. Related problems led a review committee, chaired by H. W. Lewis of the University of California, Santa Barbara, to conclude that the Reactor Safety Study greatly overestimated the precision with which the probability of a core meltdown could be assessed.[10]

Another case in point is the 1976 collapse of the Teton Dam. The Committee on Government Operations has attributed this disaster to the unwarranted confidence of engineers who were absolutely certain they had solved the many serious problems that arose during construction.[11] Indeed, in routine practice, failure probabilities are not even calculated for new dams even though about 1 in 300 fails when the reservoir is first filled. Further anecdotal evidence of overconfidence may be found in many other technical risk assessments. Some common ways in which experts may overlook or misjudge pathways to disaster include:

(1) Failure to consider the ways in which human errors can affect technological systems. Example: Due to inadequate training and control room design, operators at Three Mile Island repeatedly misdiagnosed the problems of the reactor and took inappropriate actions.

(2) Overconfidence in current scientific knowledge. Example: Use of DDT came into widespread and uncontrolled use before scientists had even considered the possibility of the side effects that today make it look like a mixed and irreversible blessing.

(3) Insensitivity to how a technological system functions

as a whole. Example: Though the respiratory risk of fossil-
fueled power plants has been recognized for some time, the
related effects of acid rains on ecosystems were largely
missed until very recently.

(4) Failure to anticipate human response to safety measures. Example: The partial protection offered by dams and
levees gives people a false sense of security and promotes
development of the flood plain. When a rare flood does exceed
the capacity of the dam, the damage may be considerably greater than if the flood plain had been unprotected. Similarly,
"better" highways, while decreasing the death toll per vehicle
mile, may increase the total number of deaths because they increase the number of miles driven.

Desire for Certainty

Every technology is a gamble of sorts and, like other
gambles, its attractiveness depends on the probability and
size of its possible gains and losses. Both scientific experiments and casual observation show that people have difficulty
thinking about and resolving the risk/benefit conflicts even
in simple gambles. One way to reduce the anxiety generated by
confronting uncertainty is to deny that uncertainty. The denial resulting from this anxiety-reducing search for certainty
thus represents an additional source of overconfidence. This
type of denial is illustrated by the case of people faced with
natural hazards, who often view their world as either perfectly safe or as predictable enough to preclude worry. Thus,
some flood victims interviewed by Kates[12] flatly denied
that floods could ever recur in their areas. Some thought
(incorrectly) that new dams and reservoirs in the area would
contain all potential floods, while others attributed previous floods to freak combinations of circumstances unlikely
to recur. Denial, of course, has its limits. Many people
feel that they cannot ignore the risks of nuclear power. For
these people, the search for certainty is best satisfied by
outlawing the risk.

Scientists and policy-makers who point out the gambles
involved in societal decisions are often resented for the
anxiety they provoke. Borch[13] noted how annoyed corporate
managers get with consultants who give them the probabilities
of possible events instead of telling them exactly what will
happen. Just before a blue-ribbon panel of scientists reported that they were 95 percent certain that cyclamates do not
cause cancer, Food and Drug Administration Commissioner, Alex-
ander Schmidt, said, "I'm looking for a clean bill of health,
not a wishy-washy, iffy answer on cyclamates".[14] Senator
Edmund Muskie has called for "one-armed" scientists who do not
respond "on the one hand, the evidence is so, but on the other

hand..." when asked about the health effects of pollutants.[15]
As Gori warns in Chapter 11, such demands may tempt scientists
to issue "certain" answers which, however convenient for regulators, are unsupportable by science.

The search for certainty is legitimate if it is done con-
sciously, if the remaining uncertainties are acknowledged rather than ignored, and if people realize the costs. If a very
high level of certainty is sought, those costs are likely to
be high. Eliminating the uncertainty may mean eliminating the
technology and foregoing its benefits. Often some risk is
inevitable. Efforts to eliminate it may only alter its form.
We must choose, for example, between the vicissitudes of nature on an unprotected flood plain and the less probable, but
potentially more catastrophic, hazards associated with dams
and levees (Coates, Chapter 3; Okrent, Chapter 13).

Analyzing Judgments of Risk

In order to be of assistance in the hazard management
process, a theory of perceived risk must explain people's extreme aversion to some hazards, their indifference to others,
and the discrepancies between these reactions and experts'
recommendations. Why, for example, do some communities react
vigorously against locating a liquid natural gas terminal in
their vicinity despite the assurances of experts that it is
safe? Why do other communities situated on flood plains and
earthquake faults or below great dams show little concern for
the experts' warnings? Such behavior is doubtless related to
how people assess the quantitative characteristics of the hazards they face. The preceding discussion of judgmental processes was designed to illuminate this aspect of perceived
risk. The studies reported below broaden the discussion to
include more qualitative components of perceived risk. They
ask, when people judge the risk inherent in a technology, are
they referring only to the (possibly misjudged) number of
people it could kill or also to other, more qualitative features of the risk it entails?

Quantifying Perceived Risk

In our first studies, we asked four different groups of
people to rate thirty different activities and technologies
according to the present risk of death from each.[16] Three
of these groups were from Eugene, Oregon; they included 30
college students, 40 members of the League of Women Voters
(LOWV), and 25 business and professional members of the "Active Club." The fourth group was composed of 15 persons selected nation-wide for their professional involvement in
risk assessment. This "expert" group included a geographer,
an environmental policy analyst, an economist, a lawyer, a

Table 10.2. Ordering of Perceived Risk for 30 Activities and Technologies[a]

	Group 1 LOWV	Group 2 Coll. Stud.	Group 3 Active Club	Group 4 Experts
Nuclear power	1	1	8	20
Motor vehicles	2	5	3	1
Handguns	3	2	1	4
Smoking	4	3	4	2
Motorcycles	5	6	2	6
Alcoholic beverages	6	7	5	3
General (private) aviation	7	15	11	12
Police work	8	8	7	17
Pesticides	9	4	15	8
Surgery	10	11	9	5
Fire fighting	11	10	6	18
Large construction	12	14	13	13
Hunting	13	18	10	23
Spray cans	14	13	23	26
Mountain climbing	15	22	12	29
Bicycles	16	24	14	15
Commercial aviation	17	16	18	16
Electric power	18	19	19	9
Swimming	19	30	17	10
Contraceptives	20	9	22	11
Skiing	21	25	16	30
X-rays	22	17	24	7
High school & college football	23	26	21	27
Railroads	24	23	20	19
Food preservatives	25	12	28	14
Food coloring	26	20	30	21
Power mowers	27	28	25	28
Prescription antibiotics	28	21	26	24
Home appliances	29	27	27	22
Vaccinations	30	29	29	25

[a]The ordering is based on the geometric mean risk ratings within each group. Rank 1 represents the most risky activity or technology.

biologist, a biochemist, and a government regulator of hazardous materials.

All these people were asked, for each of the thirty items, "to consider the risk of dying (across all U.S. society as a whole) as a consequence of this activity or technology." In order to make the evaluation task easier, each activity appeared on a 3" x 5" card. Respondents were told first to study the items individually, thinking of all the possible ways someone might die from each (e.g., fatalities from nonnuclear electricity were to include deaths resulting from the mining of coal and other energy production activities as well as electrocution; motor vehicle fatalities were to include collisions with bicycles and pedestrians). Next, they were to order the items from least to most risky and then assign numerical risk values by giving a rating of 10 to the least risky item and making the other ratings accordingly. They were also given additional suggestions, clarifications, and encouragement to do as accurate a job as possible. For example, they were told "A rating of 12 indicates that that item is 1.2 times as risky as the least risky item (i.e., 20 percent more risky). A rating of 200 means that the item is 20 times as risky as the least risky item, to which you assigned a 10 . . . " They were urged to cross-check and adjust their numbers until they believed they were right.

Table 10.2 shows how the various groups ranked the relative riskiness of these 30 activities and technologies. There were many similarities between the three groups of lay persons. For example, each group believed that motorcycles, other motor vehicles, and handguns were highly risky, and that vaccinations, home appliances, power mowers, and football were relatively safe. However, there were strong differences as well. Active Club members viewed pesticides and spray cans as relatively much safer than did the other groups. Nuclear power was rated as highest in risk by the LOWV and student groups, but only eighth by the Active Club. The students viewed contraceptives and food preservatives as riskier and swimming and mountain climbing as safer than did the other lay groups. Experts' judgments of risk differed markedly from the judgments of lay persons. The experts viewed electric power, surgery, swimming, and X-rays as more risky than the other groups, and they judged nuclear power, police work, and mountain climbing to be much less risky.

What Determines Risk Perception?

What do people mean when they say that a particular technology is quite risky? A series of additional studies was conducted to answer this question.

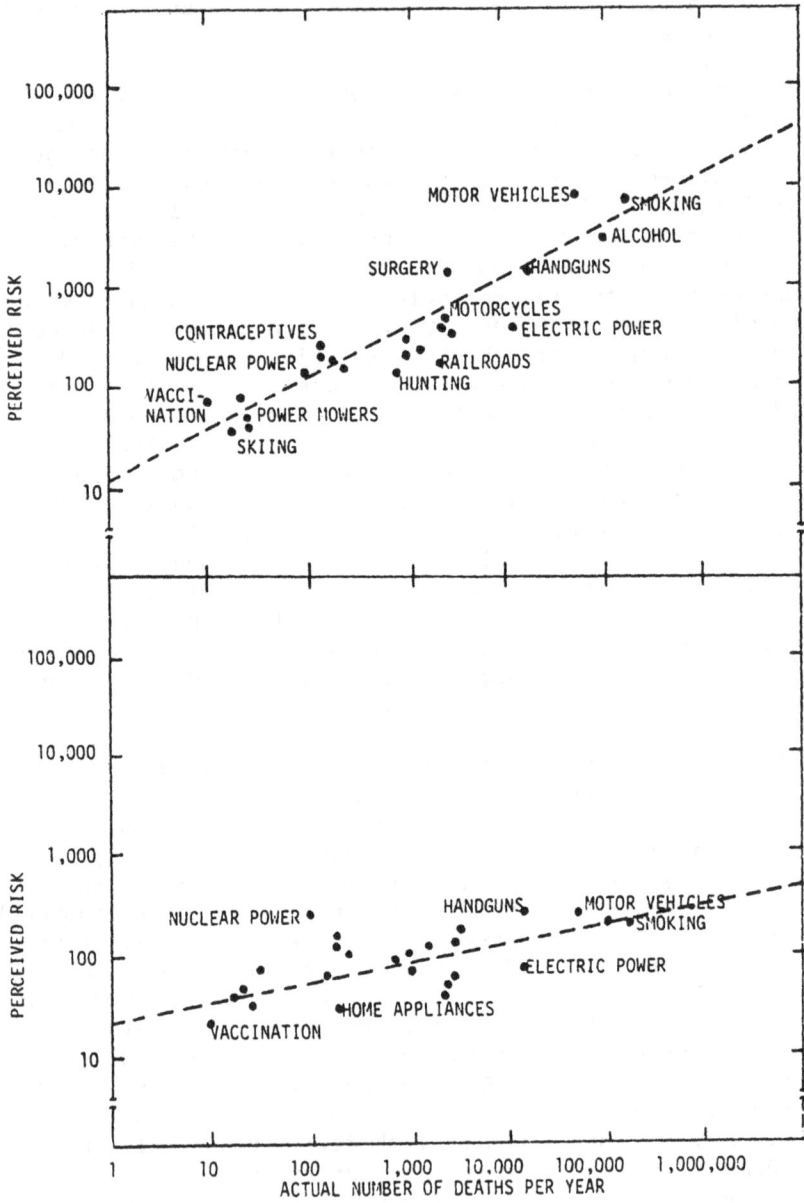

Figure 10.3. Judgments of perceived risk for experts (top) and lay people (bottom) plotted against the best technical estimates of annual fatalities for 25 technologies and activities. Each point represents the average responses of the participants. The dashed lines are the straight lines that best fit the points. The experts' risk judgments are seen to be more closely associated with annual fatality rates than are the lay judgments.

Perceived risk compared to frequency of death.   When
people judge risk, as in the previous study, are they simply
estimating frequency of death? To answer this question, we
collected the best available technical estimates of the annual
number of deaths from each of the thirty activities included
in our study. For some cases, such as commercial aviation
and handguns, there is good statistical evidence based on
counts of known victims. For other cases, such as the lethal
potential of nuclear or fossil-fuel power plants, available
estimates are based on uncertain inferences about incompletely
understood processes. For still others, such as food coloring, we could find no estimates of annual fatalities.

For the 25 cases for which we found technical estimates
for annual frequency of death, we compared these estimates
with perceived risk. Results for experts and the LOWV sample
are shown in Figure 10.3 (the results for the other lay groups
were quite similar to those from the LOWV sample). The experts' mean judgments were so closely related to the statistical or calculated frequencies that it seems reasonable to
conclude that they viewed the risk of an activity or technology as synonymous with its annual fatalities. The risk judgments of lay people, however, showed only a moderate relationship to the annual frequencies of death[17], raising the
possibility that, for them, risk may not be synonymous with
fatalities. In particular, the perceived risk from nuclear
power was disproportionately high compared to its estimated
number of fatalities.

Lay fatality estimates.   Perhaps lay people based their
risk judgments on annual fatalities but estimated their numbers inaccurately. To test this hypothesis, we asked additional groups of students and LOWV members "to estimate how
many people are likely to die in the U.S. in the next year
(if the next year is an average year) as a consequence of
these thirty activities and technologies" We asked our student and LOWV samples to consider all sources of death associated with these activities.

The mean fatality estimates of LOWV members and students
are shown in columns 2 and 3 of Table 10.3. If lay people
really equate risk with annual fatalities, one would expect
that their own estimates of annual fatalities, no matter how
inaccurate, would be very similar to their judgments of risk.
But this was not so. There was a moderate agreement between
their annual fatality estimates and their risk judgments, but
there were important exceptions. Most notably, nuclear power
had the lowest fatality estimate and the highest perceived
risk for both LOWV members and students. Overall, lay people's perceptions were no more closely related to their own

Table 10.3. Fatality Estimates and Disaster Multipliers for 30 Activities and Technologies

Activity or Technology	Technical Fatality Estimates	Geometric Mean Fatality Estimates Average Year		Geometric Mean Multiplier Disastrous Year	
		LOWV	Students	LOWV	Students
Smoking	150,000	6,900	2,400	1.9	2.0
Alcoholic beverages	100,000	12,000	2,600	1.9	1.4
Motor vehicles	50,000	28,000	10,500	1.6	1.8
Handguns	17,000	3,000	1,900	2.6	2.0
Electric power	14,000	660	500	1.9	2.4
Motorcycles	3,000	1,600	1,600	1.8	1.6
Swimming	3,000	930	370	1.6	1.7
Surgery	2,800	2,500	900	1.5	1.6
X-rays	2,300	90	40	2.7	1.6
Railroads	1,950	190	210	3.2	1.6
General (private) aviation	1,300	550	650	2.8	2.0
Large construction	1,000	400	370	2.1	1.4
Bicycles	1,000	910	420	1.8	1.4
Hunting	800	380	410	1.8	1.7
Home appliances	200	200	240	1.6	1.3
Fire fighting	195	220	390	2.3	2.2
Police work	160	460	390	2.1	1.9
Contraceptives	150	180	120	2.1	1.4
Commercial aviation	130	280	650	3.0	1.8
Nuclear power	100[a]	20	27	107.1	87.6
Mountain climbing	30	50	70	1.9	1.4
Power mowers	24	40	33	1.6	1.3
High school & college football	23	39	40	1.9	1.4
Skiing	18	55	72	1.9	1.6
Vaccinations	10	65	52	2.1	1.6
Food coloring	--[b]	38	33	3.5	1.4
Food preservatives	--[b]	61	63	3.9	1.7
Pesticides	--[b]	140	84	9.3	2.4
Prescription antibiotics	--[b]	160	290	2.3	1.6
Spray cans	--[b]	56	38	3.7	2.4

[a] Technical estimates for nuclear power were found to range between 16 and 600 annual fatalities. The geometric mean of these estimates was used here.

[b] Estimates were unavailable.

fatality estimates than they were to the technical estimates
(Figure 10.3).

These results lead us to reject the idea that lay people
wanted to equate risk with annual fatality estimates but were
inaccurate in doing so. Instead, we are led to believe that
lay people incorporate other considerations besides annual
fatalities into their concept of risk.

Some other aspects of lay people's fatality estimates
are of interest. One is that they were moderately accurate.
The relationship between the LOWV members' fatality estimates
and the best technical estimates is plotted in Figure 10.4.
The lay estimates showed the same overestimation of those
items that cause few fatalities and underestimation of those
resulting in the most fatalities that was apparent in Figure
10.2 for a different collection of hazards. Also as in Figure 10.2, the moderate overall relationship between lay and
technical estimates was marred by specific biases (e.g.,
the underestimation of fatalities associated with railroads,
X-rays, electric power, and smoking).

Disaster potential. The fact that the LOWV members and
students assigned very high risk values to nuclear power along
with very low estimates of its annual fatality rates is an
apparent contradiction. One possible explanation is that
LOWV members expected nuclear power to have a low death rate
in an average year but considered it to be a high-risk technology because of its potential for disaster.

In order to understand the role played by expectations
of disaster in determining lay people's risk judgments, we
asked these same respondents to give for each activity and
technology a number indicating how many times more deaths
would occur if next year were "particularly disastrous"
rather than average. The averages of these multipliers are
shown in Table 10.3. For most activities, people saw little
potential for disaster. For the LOWV sample all but five of
the multipliers were less than 3, and for the student sample
all but six were less than 2. The striking exception in both
cases is nuclear power, with a geometric mean disaster multiplier in the neighborhood of 100.

For any individual an estimate of the expected number of
fatalities in a disastrous year could be obtained by applying
the disaster multiplier to the estimated fatalities for an
average year. When this was done for nuclear power, almost
40 percent of the respondents expected more than 10,000 fatalities if next year were a disastrous year. More than 25 percent expected 100,000 or more fatalities. These extreme esti-

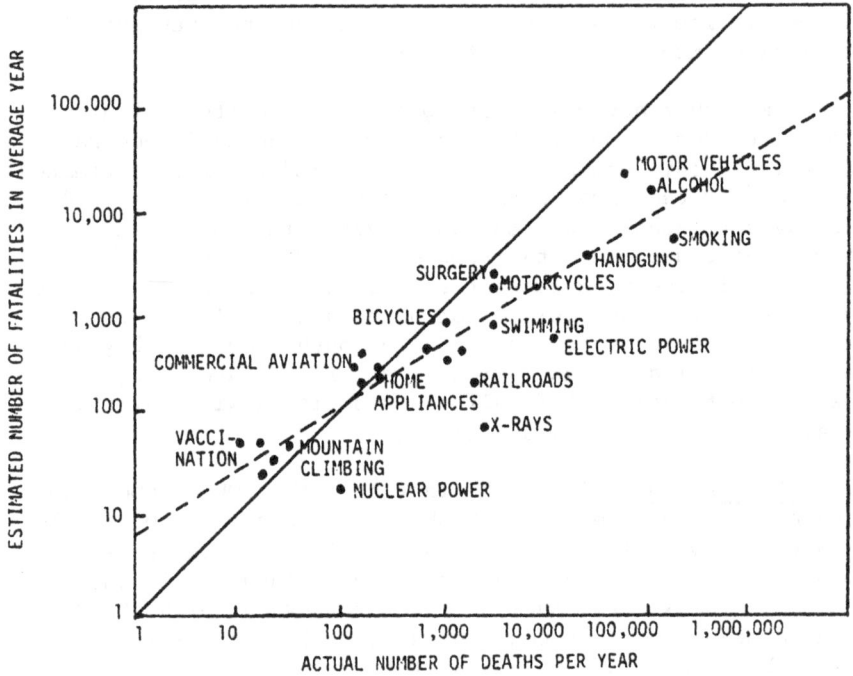

Figure 10.4. Lay people's judgments of the number of fatalities in an average year plotted against the best estimates of
annual fatalities for 25 activities and technologies. The
solid line indicates accurate judgment, while the dashed line
best fits the data points. These results have much the same
character as those shown in Figure 10.2 for a different collection of hazards. Low frequencies were overestimated and
high ones were underestimated. The overall relationship is
marred by specific biases (e.g., the underestimation of fatalities associated with railroads, X-rays, electric power, and
smoking).

mates can be contrasted with the Reactor Safety Study's conclusion that the maximum credible nuclear accident, coinci-
dent with the most unfavorable combination of weather and population density, would cause only 3,300 prompt fatalities.[9]
Furthermore, that study estimated the odds against an accident of this magnitude occurring during the next year (assuming 100 operating reactors) to be about 2,000,000:1.

Apparently, disaster potential explains much or all of
the discrepancy between the perceived risk and frequency of
death values for nuclear power. Yet, because disaster plays
only a small role in most of the thirty activities and technologies we have studied, it provides only a partial explanation of the perceived risk data.

Qualitative characteristics. Are there other determinants of risk perceptions besides frequency estimates? We
asked experts, students, LOWV members, and Active Club members to rate the thirty technologies and activities on nine
qualitative characteristics that have been hypothesized to
be important.[18] These ratings scales are described in
Table 10.4.

Mean ratings were quite similar for all four groups.
Particularly interesting was the characterization of nuclear
power, which had the dubious distinction of scoring at or
near the extreme on all of the undesirable characteristics.
Its risks were seen as involuntary, delayed, unknown, uncon-
trollable, unfamiliar, catastrophic, dread, and fatal. This
contrasted sharply with the characterizations of non-nuclear
electric power and another radiation technology, X-rays.
Electric power and X-rays were both judged more voluntary,
less certain to be fatal, less catastrophic, less dreaded,
more familiar, and less risky than nuclear power (see Figure 10.5).

Across all 30 hazards, ratings of dread and of the likelihood of a mishap's being fatal were closely related to lay
judgments of risk. In fact, the risk judgments of the LOWV
and student groups could be predicted almost perfectly from
ratings of dread and lethality and the subjective fatality
estimates for normal and disastrous years.[19] Experts'
judgments of risk were not related to any of the nine risk
characteristics.[20]

Many pairs of risk characteristics tended to be correlated with each other across the 30 activities and technologies. For example, risks faced voluntarily were typically
judged well known and controllable. These interrelations
were sufficiently high to suggest that all the ratings could
be explained in terms of a few basic dimensions of risk. In

Table 10.4. Risk characteristics rated by LOWV members, Active Club members, students, and experts

---

Voluntariness of risk
Do people face this risk voluntarily? If some of the risks are voluntarily undertaken and some are not, mark an appropriate spot towards the center of the scale.

risk assumed   *risk assumed*
voluntarily   1 2 3 4 5 6 7   *involuntarily*

Immediacy of effect
To what extent is the risk of death immediate--or is death likely to occur at some later time?

effect immediate 1 2 3 4 5 6 7   *effect delayed*

Knowledge about risk
To what extent are the risks known precisely by the persons who are exposed to those risks?

risk level known   *risk level not*
precisely   1 2 3 4 5 6 7   *known*

To what extent are the risks known to science?

risk level known   *risk level not*
precisely   1 2 3 4 5 6 7   *known*

Control over risk
If you are exposed to the risk, to what extent can you, by personal skill or diligence, avoid death?

personal risk can't   *personal risk can*
be controlled   1 2 3 4 5 6 7 *be controlled*

Newness
Is this risk new and novel or old and familiar?

new   1 2 3 4 5 6 7   *old*

Chronic/Catastrophic
Is this a risk that people have learned to live with and can think about reasonably calmly, or is it one that people have great dread for--on the level of a gut reaction?

common   1 2 3 4 5 6 7   *dread*

Severity of consequences
When the risk from the activity is realized in the form of a mishap or illness, how likely is it that the consequence will be fatal?

certain not to be   *certain to be*
fatal   1 2 3 4 5 6 7   *fatal*

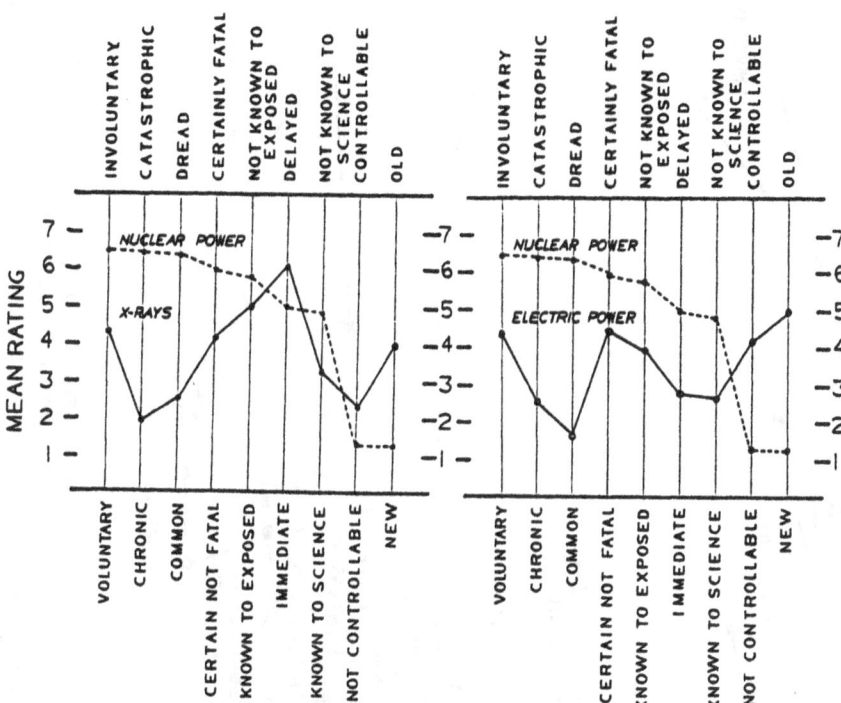

Figure 10.5.　Qualitative characteristics of perceived risk for nuclear power and related technologies. In the right-hand diagram, risk profiles for nuclear power and non-nuclear electric power are compared. In the left-hand diagram, nuclear power and X-rays are compared. Each profile consists of nine dimensions rated on a seven-point scale. The instructions that elicited these responses are reproduced in Table 10.4. The perceived qualities of nuclear power are dramatically different from the comparison technologies. The source of this data is the LOWV sample studied by Fischhoff et al.[16]

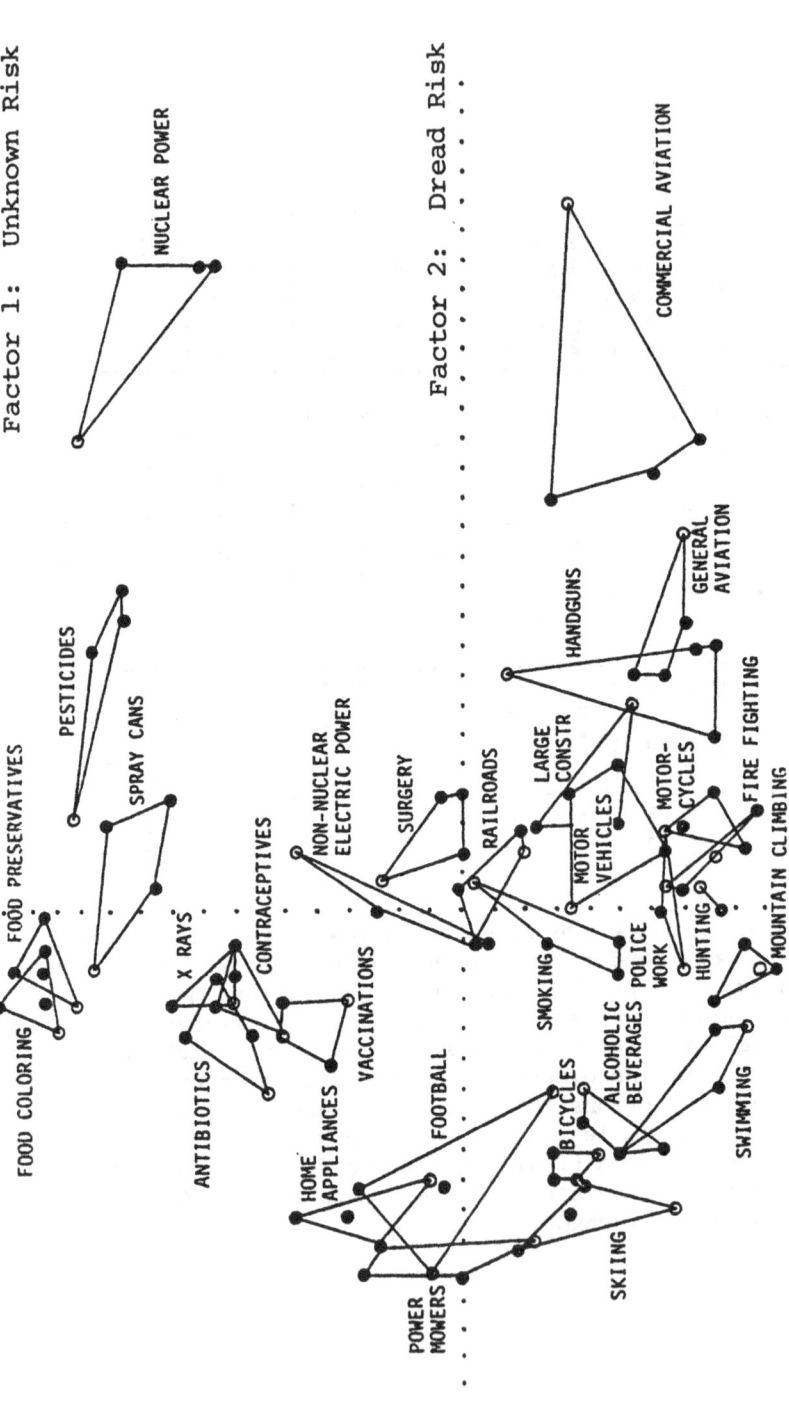

Figure 10.6. Location of 30 hazards within the two-factor space obtained from LOWV, student, Active Club and expert groups. Connected lines join or enclose the loci of four group points for each hazard. Open circles represent data from the expert group. Unattached points represent groups that fall within the triangle created by the other three groups.

order to identify such dimensions, we conducted a factor analysis of the correlations from each group (principal components analysis with varimax rotation to simple structure). We found that the nine characteristics could be represented by two underlying factors which appeared to be the same for each group. Figure 10.6 illustrates the factor scores for each hazard within the common space. Hazards at the high end of the vertical dimension or factor (e.g., food coloring, pesticides) tended to be new, unknown, involuntary, and delayed in their effects. Hazards at the other extreme of this factor (e.g., mountain climbing, swimming) had the opposite characteristics. High (right-hand) scores on the horizontal factor (e.g., nuclear power, commercial aviation) were associated with events whose consequences were seen as certain to be fatal, often for large numbers of people, should something go wrong. Hazards low on this factor (e.g., power mowers, football) were seen as causing injuries, rather than fatalities, to single individuals. We have labeled the vertical factor as "Unknown Risk" and the horizontal factor as "Dread Risk." In sum, even though the four groups had somewhat different perceptions of the riskiness of the various hazards (Table 10.2), they tended to characterize these hazards similarly.

Judged seriousness of death. In a further attempt to improve our understanding of perceived risk, we examined the hypothesis that some hazards are feared more than others because the deaths they produce are much "worse" than deaths from other activities. We thought, for example, that deaths from risks imposed involuntarily, from risks not under one's control, or from hazards that are particularly dreaded might be given greater weight in determining people's perceptions of risk.

When we asked students and LOWV members to judge the relative "seriousness" of a death from each of the thirty activities and technologies, however, the differences were slight. The most serious forms of death (from nuclear power and handguns) were judged to be only about two to four times worse than the least serious forms of death (from alcoholic beverages and smoking). Furthermore, across all thirty activities, judged seriousness of death was not closely related to perceived risk of death.

Reconciling Divergent Opinions

Our data show that experts and lay people have quite different perceptions about how risky certain technologies are. It would be comforting to believe that these divergent risk judgments would be responsive to new evidence so that, as information accumulates, perceptions would converge towards one "appropriate" view. Unfortunately, this is not likely to be the case. As noted earlier in our discussion of availability, risk perception is derived in part from fundamental modes of thought that lead people to rely on fallible indicators such as memorability and imaginability.

Furthermore, a great deal of research indicates that people's beliefs change slowly and are extraordinarily persistent in the face of contrary evidence.[21] Once formed, initial impressions tend to structure the way that subsequent evidence is interpreted. New evidence appears reliable and informative if it is consistent with one's initial belief; contrary evidence is dismissed as unreliable, erroneous, or unrepresentative. Thus, depending on one's predispositions, intense effort to reduce a hazard may be interpreted to mean either that the risks are great or that the technologists are responsive to the public's concerns. Likewise, opponents of a technology may view minor mishaps as near catastrophes and dismiss the contrary opinions of experts as biased by vested interests.

From a statistical standpoint, convincing people that the catastrophe they fear is extremely unlikely is difficult under the best conditions. Any mishap could be seen as proof of high risk, whereas demonstrating safety would require a massive amount of evidence.[22] Nelkin's case history of a nuclear siting controversy[23] provides a good example of the inability of technical arguments to change opinions. In that debate each side capitalized on technical ambiguities in ways that reinforced its own position.

The Fallibility of Judgment

Our examination of risk perception leads us to the following conclusions:

(1) Cognitive limitations, coupled with the anxieties generated by facing life as a gamble, cause uncertainty to be denied, risks to be distorted, and statements of fact to be believed with unwarranted confidence.

(2) Perceived risk is influenced (and sometimes biased) by the imaginability and memorability of the hazard. People may, therefore, not have valid perceptions even for familiar risks.

(3) Our experts' risk perceptions correspond closely to statistical frequencies of death. Lay people's risk perceptions were based in part upon frequencies of death, but there were some striking discrepancies. It appears that for lay people, the concept of risk includes qualitative aspects such as dread and the likelihood of a mishap's being fatal. Lay

people's risk perceptions were also affected by catastrophic potential.

Disagreements about risk should not be expected to evaporate in the presence of "evidence." Definitive evidence, particularly about rare hazards, is difficult to obtain. Weaker information is likely to be interpreted in a way that reinforces existing beliefs.

The significance of these results hinges upon one's acceptance of our assumption that subjective judgments are central to the hazard management process. Our conclusions mean little if one can assume that there are analytical tools which can be used to assess most risks in a mechanical fashion and that all decision makers have perfect information and the know-how to use it properly. These results gain in importance to the extent that one believes, as we do, that expertise involves a large component of judgment, that the facts are not all in (or obtainable) regarding many important hazards, that people are often poorly informed or misinformed, and that they respond not just to numbers but also to qualitative aspects of hazards.

Whatever role judgment plays, its products should be treated with caution. Research not only demonstrates that judgment is fallible, but it shows that the degree of fallibility is often surprisingly great and that faulty beliefs may be held with great confidence.

When it can be shown that even well-informed lay people have difficulty judging risks accurately, it is tempting to conclude that the public should be removed from the hazard-management process. The political ramifications of such a transfer of power to a technical elite are obvious. Indeed, it seems doubtful that such a massive disenfranchisement is feasible in any democratic society.

Furthermore, this transfer of decision-making would seem to be misguided. For one thing, we have no assurance that experts' judgments are immune to biases once they are forced to go beyond their precise knowledge and rely upon their judgment. Although judgmental biases have most often been demonstrated with lay people, there is evidence that the cognitive functioning of experts is basically like that of everyone else.

In addition, in many if not most cases, effective hazard management requires the cooperation of a large body of lay people. These people must agree to do without some things and accept substitutes for others; they must vote sensibly on ballot measures and for legislators who will serve them as surrogate hazard managers; they must obey safety rules and use the legal system responsibly. Even if the experts were much better judges of risk than lay people, giving experts an exclusive franchise on hazard management would involve substituting short-term efficiency for the long-term effort needed to create an informed citizenry.

For those of us who are not experts, these findings pose an important series of challenges: to be better informed, to rely less on unexamined or unsupported judgments, to be aware of the qualitative aspects that strongly condition risk judgments, and to be open to new evidence that may alter our current risk perceptions.

For the experts, our findings pose what may be a more difficult challenge: to recognize their own cognitive limitations, to temper their assessments of risk with the important qualitative aspects of risk that influence the responses of lay people, and somehow to create ways in which these considerations can find expression in hazard management without, in the process, creating more heat than light.

Acknowledgment

The authors wish to express their appreciation to Christoph Hohenemser, Roger Kasperson, and Robert Kates for their many helpful comments and suggestions. This work was supported by the National Science Foundation under Grant ENV77-15332 to Perceptronics, Inc. Any opinions, findings and conclusions, or recommendations expressed herein are those of the authors and do not necessarily reflect the views of the National Science Foundation.

References and Notes

1. We have not attempted here to review all the important research in this area. Interested readers should see C.H. Green, "Risk: Attitudes and Beliefs," in *Behavior in Fires*, ed. D.V. Canter (Wiley, New York, in press); R.W. Kates, *Risk Assessment of Environmental Hazard* (Wiley, New York, 1978); and H.J. Otway, D. Maurer, and K. Thomas, "Nuclear Power: The Question of Public Acceptance," *Futures*, 10 (April, 1978), 109-118.

2. A. Tversky and D. Kahneman, "Judgment under Uncertainty: Heuristics and Biases," *Science*, 185 (1974), 1124-1131.

3. A. Tversky and D. Kahneman, "Availability: A Heuristic

for Judging Frequency and Probability," *Cognitive Psychology*, 4 (1973), 207-232.

4. S. Lichtenstein, P. Slovic, B. Fischhoff, M. Layman, and B. Combs, "Judged Frequency of Lethal Events." *Journal of Experimental Psychology: Human Learning and Memory*, 4 (1978), 551-578.

5. B. Fischhoff, P. Slovic, and S. Lichtenstein, "Fault Trees: Sensitivity of Estimated Failure Probabilities to Problem Representation," *Journal of Experimental Psychology: Human Perception and Performance*, 4 (1978), 342-355.

6. B. Fischhoff, P. Slovic, and S. Lichtenstein, "Knowing with Certainty: The Appropriateness of Extreme Confidence," *Journal of Experimental Psychology: Human Perception and Performance*, 3 (1977), 552-564.

7. S. Lichtenstein, B. Fischhoff, and L.D. Phillips, "Calibration of Probabilities: The State of the Art," *Decision Making and Change in Human Affairs*, ed. H. Jungermann and G. de Zeeuw (D. Reidel, Dordrecht, The Netherlands, 1977).

8. M. Hynes and E. Vanmarcke, "Reliability of Embankment Performance Predictions," Proceedings of the ASCE Engineering Mechanics Division Specialty Conference (University of Waterloo Press, Waterloo, Ontario, 1976).

9. U.S. Nuclear Regulatory Commission, *Reactor Safety Study: An Assessment of Accident Risks in U.S. Commercial Nuclear Power Plants* (WASH-1400, NUREG-75/014); Nuclear Regulatory Commission, Washington, 1975.

10. U.S. Nuclear Regulatory Commission, Risk Assessment Review Group, *Risk Assessment Review Group Report to the U.S. Nuclear Regulatory Commission* (NUREG/CR-0400; Nuclear Regulatory Commission, Washington, 1978).

11. U.S. Congress, House, Committee on Government Operations, Subcommittee on Conservation, Energy, and Natural Resources, *Teton Dam Disaster: Hearings...* (94th Cong., 2d sess.; Government Printing Office, Washington, 1976).

12. R.W. Kates, *Hazard and Choice Perception in Flood Plain Management* (Research Paper, 78; Department of Geography, University of Chicago, Chicago, 1962).

13. K. Borch, *The Economics of Uncertainty* (Princeton University Press, Princeton, N.J., 1968).

14. "Doubts Linger on Cyclamate Risks," *Eugene Register-Guard,* 14 January 1976.

15. E.E. David, "One-Armed Scientists?," *Science,* 189 (1975) 891.

16. B. Fischhoff, P. Slovic, S. Lichtenstein, S. Read, and B. Combs, "How Safe is Safe Enough? A Psychometric Study of Attitudes towards Technological Risks and Benefits," *Policy Sciences,* 8 (1978), 127-152; P. Slovic, B. Fischhoff, and S. Lichtenstein, "Expressed Preferences," unpublished manuscript (Decision Research, Eugene, Oregon, 1978).

17. The correlations between perceived risk and the annual frequencies of death were .92 for the experts and .62, .50, and .56 for the League of Women Voters, students, and Active Club samples, respectively.

18. W. Lowrance, *Of Acceptable Risk* (William Kaufman, Los Altos, California, 1976).

19. The multiple correlation between the risk judgments of the LOWV members and students and a linear combination of their fatality estimates, disaster multipliers, dread ratings, and severity ratings was .95.

20. A secondary finding was that both experts and lay persons believed that the risks from most of the activities were better known to science than to the individuals at risk. The experts believed that the discrepancy in knowledge was particularly great for vaccinations, X-rays, antibiotics, alcohol, and home appliances. The only activities whose risks were judged better known to those exposed were mountain climbing, fire fighting, hunting, skiing, and police work.

21. L. Ross, "The Intuitive Psychologist and His Shortcomings," *Advances in Social Psychology,* ed. L. Berkowitz (Academic Press, New York, 1977).

22. A.E. Green and A.J. Bourne, *Reliability Technology* (Wiley Interscience, New York, 1972).

23. D. Nelkin, "The Role of Experts on a Nuclear Siting Controversy," *Bulletin of the Atomic Scientists,* 30 (1974), 29-36.

Gio Batta Gori

11. The Regulation of Carcinogenic Hazards

In earlier times economic development, an ethical imperative in the Western world, took precedence over concern for individual or collective safety. Rational audacity is a distinct human characteristic, and many have argued that too much regard for safety goes against human nature and happiness. But today new dangers of chronic illness and disability give new dimensions to fear, and there is a clear need to define better the odds involved.

Carcinogens have been recognized since Percival Pott some 200 years ago,[1] and regulatory initiatives have flourished for the last three decades, as anxieties gradually shifted from vanishing acute infections to chronic diseases made more prevalent by longevity gains,[2] and in the wake of concern over ubiquitous industrialization.

The need for regulation of carcinogens in developed societies is undisputed, but the premises and practices of such regulation are not. In the United States we are experiencing a transitional period that reflects general trends of social evolution; regulatory agencies, which in accord with centuries-old traditions have been allowed to wield quasi-autonomous normative powers, find themselves increasingly at odds with expanding demands for due process in the resolution of uncertain perceptions and conflicting values.

In many countries the balance of tolerable risks and benefits is the central regulatory issue,[3] but in the United States regulation was born of the social dialectic of the turn of the century and retains some of the romantic intransigence of that movement. American laws do not give explicit directions for considering risk/benefit factors, and

Reprinted by permission from Science, vol. 208 (18 April 1980), pp. 256-261. Copyright 1980 by the American Association for the Advancement of Science.

economic arguments usually are contested with the implicit assertion that health has a supreme value whose economic costs need not, should not, be measured.

Undeniably, an individual life has transcendent value. But excessive regulation hampers technological development and thereby denies its fruits to the poor in our own society and elsewhere in the world. Public policy is fundamentally an economic exercise. It cannot evade the balancing of risks and benefits without incurring gross inequities.

Gradually, loftier views are giving way to a realism that expects regulation to improve the quality of life for the living, not merely to extend life expectancy. When it is accepted that absolute safety is not a reasonable goal, it becomes the business of regulation to define tolerable levels of risk; to this end, explicit procedures for benefit assessment need to be introduced into regulatory statutes. Also, revisions of current practices appear to be in order for a consistent approach to the determination of risk from potential carcinogens.

In general, normative regulation has relied on the definition of standards, usually of empirical origin but then upgraded as use and newly acquired information suggest: the standards define reference compounds, testing procedures, process flows, as well as tolerable or permissible doses and levels.

In the late 1950s, in the heat of public concern over food additives and new pharmaceuticals, U.S. legislators sought the advice of science in the definition of carcinogen standards. Unfortunately, at that time understanding in this field was too problematic to produce even a suggestion of standards. Legislators were left with the alternatives of either intransigent policies, such as the Delaney amendment adopted at that time, or vague statements of intent, which have intrigued dialecticians ever since. It now appears that these legislative precedents have been largely responsible for the polarization and ambiguity that plague the regulation of carcinogens in this country and for a climate of opinion that has discouraged intermediate solutions in favor of all-or-none pronouncements.

The public would be surprised to note that different potential hazards, documented with comparable scientific methods and data, are regulated by widely different criteria. The explanation is simple. There is an implicit necessity to tolerate certain conditions where intransigent regulation would mean a drastic alteration of traditional life-styles. There are many examples: exposure to sunlight; ingestion of

fats, proteins, and excessive calories, or of foods containing natural and apparently unavoidable potential carcinogens
such as aflatoxins; and the paradox of potential risk from
the ingestion of our own saliva, at times very rich in ni-
trites, precursors of carcinogenic nitrosamines.

The need to tolerate such hazards has not been seriously challenged even by the most ardent proponents of regulation. But the absence of explicit statutory rules for risk
and benefit assessment has led to a situation where each substance is considered separately, and widely disparate outcomes are influenced more by adversary emotions than by real
values. Testing procedures have proliferated and become more
complicated, logistic resources have been virtually exhaust-
ed, and testing as now envisioned may be precluded for the
majority of environmentally significant substances.

Moreover, intrinsic uncertainties in current procedures
make it impossible to prove safety beyond doubt; on this
basis prognostications of doom have flourished. Such prophe-
cies fly in the face of a little-publicized circumstance,
namely that in the United States and other advanced countries
age-adjusted cancer rates in general have remained nearly
stationary or have declined over the past several decades,
except for some few cancers of recognized etiology.[1,4]

Yet, because regulation is an essential safeguard of
civilized living, we must resolve the conflict between the
need to improve living standards and the need to preserve
health and the natural environment. We must find more rational and defensible regulatory options.

Of prime importance will be a reevaluation of the scientific framework for the appraisal of carcinogenic risks.
For this purpose society needs to depend on the objectivity
of scientists, free of political pressures in experimental
choices, design, and interpretation.[5] Today such pressures
are not absent; scientists have often been forced to produce
clear-cut statements that, however convenient for the regulator, may not have scientific justification.

Many of the debates on regulation of carcinogens have
relied on the notion that animal tests can provide meaningful
data for extrapolation of human risk (see, for example, Chapter 15), but evidence reviewed in this article suggests that
this notion should be modified.

What Is a Carcinogen?

It is commonly observed that higher organisms are naturally affected by tumors. Because the information to deter-

mine their origins is lacking, it has been customary to define as baseline the natural rate of incidence of tumors in
populations that have not been disturbed by known challenges.
Thus, the current definition of carcinogens refers to insults
that increase the incidence of all tumors or of certain tumors or that shorten their customary time of appearance.[6-8]
This teleologic definition identifies the biologic consequences of carcinogens but not the mechanisms of action involved, which at present are still unknown.[9] Under such definition, overcrowding, noise, and circadian and other
stresses[10] could come to be defined as carcinogens even
though they may be merely modulating factors in the assay
system used.

It has been argued that for regulatory purposes it does
not matter that the definition is only teleologic, because
it is the final outcome, increased cancer incidence, that
regulation seeks to control. This would be true if simple
and reproducible interactions existed within the modulating
factors and carcinogens that determine response in a given
assay, but even in those instances where semi-quantitative
outcomes can be experimentally reproduced, some artificial
contrivance is necessary, such as the use of compounds that
are strong carcinogens for the species or strain selected--
but not necessarily for others--or the use of the maximum
doses the test animals can tolerate. Both instances represent limiting situations, where the carcinogenic insult is
artificially made to overpower other factors.

It is well known that general toxicity and carcinogenicity do not go hand in hand and that they vary from species
to species and with the method of administration or intake;
therefore, maximum tolerated doses--generally the highest intake that an animal can sustain for its lifetime without significant signs of acute or subchronic toxicity--will result
in widely disparate testing levels, quite at odds with real-
life conditions.

It is apparent that the current definition of carcinogens confines the validity of data to a specific experiment,
restricts the opportunities for generalization, and makes it
difficult to distinguish between direct carcinogens and modifying factors.

The Carcinogenesis Process

Essentially nothing is known of the ultimate molecular
events that determine the transformation of normal cells into cancer cells.[11 (p.7),12] Tumors can originate spontaneously,
thus suggesting a natural stability of the cell. Whether
this is truly an intrinsic phenomenon has yet to be finally

settled; more often transformation is observed after the application of an insult external to the cell.

Today it is generally believed that a carcinogen entering an animal may undergo various metabolic manipulations before reaching a target cell; there it may find various conditions of susceptibility, resistance, competition, or repair, and may be subject to additional modification before finding molecular receptors to determine, directly or not, transformation and the eventual appearance of a cancer cell,[13] either as a single hit or after cumulation of progressive insults and damage. Not all cancer cells developed through this process have the opportunity of progressing to overt disease, because natural defenses may suppress the onset of asymptomatic cancer.

It is estimated[11](p.11) that the potency of carcinogens in animal experiments can vary by a factor of 10^7. Human exposure to environmental insults can vary by a factor of 10^9.[14] If one were to add the attenuations occurring between exposure and actual intake of a compound, it is conceivable that the range of effectiveness of a carcinogen could vary by several orders of magnitude, and the overall probability of a given molecule's being effective could become very small indeed.

As the outcome of cancer appears to be determined by the balance against the effectiveness of a single insult entity and the frequency of available entities, the regulatory process attempts to identify the quantity of insult to which man can be exposed without an unacceptable chance of developing cancer.

Because it would be unethical to conduct prospective testing in humans, animal experimentation has been used as an alternative, with the implication that carcinogenesis data from animal experiments can be translated to human conditions.

Testing for Carcinogens in Animals

Using animals to test carcinogens has its roots in basic research where the principal concern was not and is not assessment of real-life risk but the study of the phenomenon of carcinogenesis. For that purpose negative results are unfruitful and there is understandable preoccupation with increasing the odds of inducing cancer. Hence, the practice developed of using maximum tolerated doses and this in species and strains chosen for their susceptibility.

So it happens that current guidelines for carcinogen bioassay are replete with precise directions for the control of room temperature, air changes, humidity, and other easily controlled conditions but often suggest the introduction of deliberate bias into the experimental design,[7,15,16] for example: "Both sexes of each of at least two species of animals should be used in the test throughout their lifespan. In most cases these species would be rats and mice. Hamsters and dogs might be suitable, but guinea pigs, for example, appear to be resistant to some known carcinogens"[17(p.8)] and presumably should not be used; and again "...considerations in selecting the proper species and strains should include... sensitivity to tumor induction...,"[15(p.4)] or "Generally such decisions have been made on the basis of the most sensitive species tested."[16(p.3)]

Also, testing guidelines in general prescribe feeding whenever that is more convenient than other modes of administration. Agents that in reality are absorbed by respiration or through the skin are thus subjected to abnormal metabolic processing and may inpinge on cellular and organ systems that are not their natural targets; in general, feeding results in higher maximum-tolerated doses, because other routes offer less protected, more direct and rapid access to receptors that determine acute toxicity. Such recommendations are justified in the design of a research experiment of self-contained validity, but are difficult to reconcile with the need to obtain results of general value, particularly when other powerful obstacles exist.

For instance, it seems reasonable to conclude that, because the probability of developing cancer from natural exposure is related to the number of cells present in an animal and to the duration of its life, aged mice should have natural cancer incidences much lower than aged humans. In fact they have comparable incidences, which suggests that mice could be from 3×10^4 to 10^9 times more cancer-prone than humans.[18,19] The compelling power of such an argument cannot be dismissed simply because of its simplicity.

Further bias derives from the usual prescription of nearly toxic "maximum tolerated" doses. Although these may not appreciably affect the animal's visible condition during the experiment, they are known to cause metabolic overloads that may unpredictably promote or retard a carcinogenic process, with outcomes that differ from species to species. Disturbing questions on this issue have been raised by many reports and studies[11,16,17,20-26] but usually go unanswered when actual recommendations for testing are made.[7,11,15-17,20,23,27]

Diet is likely to be a major source of experimental var-

iation. Early observations[28] on the effect of caloric intake, of dietary fat and protein, have been followed by an
even broader appreciation of the enzyme inducers and toxi-
cants that may act, for example, on the immune system.[29-31]
Moreover, when the agent being tested is a promoter, the
presence of carcinogenic contaminants in the diet may erron-
eously result in its classification as a carcinogen, with
outcomes that may vary from species to species and from diet
to diet.

Concern with dietary disturbances has been voiced in
many reports: "...some natural constituents of the diet or
even an essential nutrient, such as selenium, may constitute a carcinogenic risk. Clearly these substances cannot
be completely excluded from the diet,"[17(p.5)] or "What can
be the significance of the incidence of...tumors in susceptible strains when one is not certain about the presence of
carcinogenic contaminants in the diet on which animals have
been maintained?"[23] The need to control diets, although
frequently recognized[23(p.427)] is still an unresolved problem
in the official guidelines for carcinogenesis experiments.[15]

Another source of difficulty is the translation of animal pathology into terms of human significance. Individual
agents can produce different tumors in different species or
only in certain species, thereby implying a variety of organ-
otropisms probably related to widely different metabolic conditions and homeostatic mechanisms of cell proliferation and
repair in different species. Hepatomas, for instance, are
very frequent in rodent tests but remarkably rare in man,
and the oncogenic viruses commonly infesting small rodents
may be one reason for the unusually high frequency of lym-
phomas in these animals. The difficulties of comparison
under these conditions are further complicated when tumors
arise from tissues of different embryologic origin in different species. The biologic implication is that different
agents may be carcinogenic for certain species or particular
organs but relatively harmless for others, for reasons that
are not yet apparent to science.

Kraybill[29] lists a number of other sources of quantitative uncertainty in animal testing, including inappropriate
routes of administration, enhancement of susceptibility by
deliberate immunosuppression or induced hormonal action, contaminants in the agents being tested, accumulation of a burden of the agent or other uncontrolled compounds in certain
tissues, the theoretical and practical difficulties in matching duration of exposure in man and animals, and differences
in the time required for tumor formation. One could add recent findings on the quantitative disturbances caused by var-

ious environmental and chemical stresses,[10,32] the use of rodent strains contaminated with endemic oncogenic viruses and of selected or inbred strains,[23] and the effects of transient infectious contaminants.[33]

In general, one can only conclude that current guidelines for the testing of carcinogens frequently introduce deliberate bias in order to enhance the probability of a positive response and that they ignore a number of sources of variability that cannot be controlled or are difficult to control with available technology. Under current testing a carcinogen may go undetected in a particular test, but just as likely a positive result may be valid only for the particular species and test conditions utilized; current science cannot predict or explain the outcome.

Current Methods of Assessing Human Risk

In present regulation, data from animal tests have been used, first, to define a particular hazard as a carcinogen--according to the general definition discussed above--and, second, as a basis for extrapolating to presumed conditions of human exposure.

The first use has been challenged because maximum tolerated doses may inhibit the appearance of tumors, as in the case of vinyl chloride[25] and other compounds, and because the metabolic overload created by such doses is likely to derange normal homeostasis and create physiologic conditions with no real-life counterpart.[16,17,21,23,24,26,30,34,35] Since false negatives are difficult to count, however, popular convention has it that current practices enhance the probability of detecting carcinogens in animals and prudence dictates that those detected should be deemed potential carcinogens for man. Whereas the latter argument may be defensible,[36] it does not provide scientific justification for the codified practice of using maximum tolerated doses.[37,38]

Of course, identification of an animal carcinogen is only a first step in the regulatory process, unless it happens to come under the provisions of the Delaney clause.[39] In that case the regulatory verdict is unequivocal, because the law states that any substance against which there is evidence must be banned. This legislation has endured for over 20 years, but lately increased analytical sophistication and expanded testing activities have begun to raise public opinion in favor of a less intransigent approach.

Congressional action, and the temporary suspension of

the Delaney requirement, in the case of saccharin is the most recent example of this trend.[40] The real-life question in this case is whether the risk from exposure to artificial sweeteners is balanced by risks that users would incur without them from diabetes, excessive calorie intake, dental caries, and so on, or simply by hedonistic rewards.

Similar questions are likely to become a major issue of regulatory action in the near future, influenced by emerging attitudes toward no-effect thresholds and toward the limitations of animal test data.

The issue of no-effect thresholds will inevitably assume importance as smaller and smaller quantities of potentially hazardous compounds become identifiable through advances in chemical methods. Up to now, the probable occurrence of thresholds has usually been ignored, and some regulatory guidelines specifically prevent considering them.[7,15] Such an attitude largely results from avoiding the distinction between the practical and the theoretical. Difficulties in conceiving or measuring thresholds in cellular and molecular contexts have been taken as reason to question the reality of those practical levels below which adverse effects cannot be measured epidemiologically.

Tolerable limits of exposure (TLVs) are a common concept in regulation, and while it is true that epidemiologic definition of practical thresholds has been difficult except in rare instances,[41] their presence is suggested by much evidence[11(p.10),42,43] which parallels universally accepted concepts in chemistry, physiology, and pharmacology. Deliberate laboratory and epidemiologic studies on this problem could supply information of direct significance to regulation.

Regarding the use of animal test data for human risk assessment, severe obstacles were recognized very early in several documents,[6,7,11,15,17,20-24,27,37-39] but the logical conclusions were not drawn; experimental practices quite valid in a basic research setting were adopted for regulatory purposes without a critical analysis of their limitations.

Once these practices were established, support was sought for them in several biometric models specifically developed to attempt a generalized quantification of human risk.[7,44,45] These statistical exercises would be justified if the animal data used in their elaboration reflected generalized human risk conditions, but they do not; nor is there a basis for deciding in which direction their results should be adjusted. The situation was recognized by an expert panel of advisers to FDA[23(p.433)] a few years ago: "...it would be

imprudent to place excessive reliance on mathematical sleight
of hand, particularly when the dose-response curves used
are largely empirical descriptions, lacking any theoretical,
physical or chemical basis." Apparently statisticians have
prudently competed with each other to produce methods that
would give the most conservative estimates, as the same FDA
panel of experts noted: "Although it is possible in principle to estimate 'safe' levels of carcinogens, uncertainties involved in the downward extrapolation from test results
will usually result in permissible levels that are the practical equivalent of zero."[23](p.435) In a general appraisal
of current procedures for human risk determination, a compelling statement has been recently advanced by Kraybill:
"...the [carcinogenic] response...is mediated and limited
by certain biochemical, metabolic, and pharmacokinetic relationships. Such boundaries must not be exceeded in biological testing and assessment of carcinogens, lest irrecon-
cilable implications are left with the scientific community
and the public, which result, in the long run, in a waste of
national resources in the interest of public health."[29] The
conclusion is that past and current testing practices do not
yield quantitative information about conditions of human
risk, and that biometric sophistication does not overcome
the limitations of these data.

An impasse is being felt in debates about regulations,[46] also reflected in confusion and contradiction at
the international level[3] as regulatory guidelines are elab-
orated by several agencies empowered by recent statutory
mandates.[15,47] Most of these attempts are based on the
traditional assumption that animal tests allow reliable
and generalized quantification of real-life carcinogenic
risk for man.

Over the last decades it has been fashionable to contrast the forthright simplicity of science with the apparent
looseness of political debate. Scientific solutions are
implicitly expected for many social difficulties; but science can draw valid conclusions only on the basis of proven
theories, controlled methods, and consistent results, none
of which are yet available in carcinogen testing. Ought
scientists to countenance the use of inconsistent data even
for such worthy causes as human health and a wholesome environment?[48,49] This question becomes yet more embarrassing
if one considers that better safety might be achieved, with
greater fairness, by an approach that explicitly recognized
the sociopolitical nature of regulation and resisted the
temptation to force arguments under scientific disguise.

Future Regulatory Directions

Discouraging as it may seem, it is not plausible that animal carcinogenesis experiments can be improved to the point where quantitative generalizations about human risk can be drawn from them. A multitude of disturbing variables is involved. There are difficulties from a logistic and a design point of view. Even the expedient of large experiments is now regarded as an improbable solution, because background noise and sources of disturbance will increase with the number of animals. Nor will current proposals of more complicated testing procedures provide a solution,[50] because the real issue is the fundamental biologic difficulty of resolving the inconsistency of chronic response in different species.

In vitro tests remain a possibility that is probably several years from practical application.[51]

With current procedures, the assay capacity in the United States is limited to a few hundred compounds a year at best; the backlog of compounds that need to be tested and the new compounds that industry would like to have tested amount every year to several tens of thousands of individual items. The new Toxic Substance Control Act alone[52] is likely to create a crisis that could only be resolved by adopting new regulatory policies, expanding resources, and simplifying testing requirements.

The crisis would be exacerbated by the continuing pressures of a consumer society that is also environment-conscious. These could swell the outcry over what appears as an exorbitant or impossible regulatory burden, force the mitigation of current requirements for testing,[34,53,54] and weaken enforcement of statutes.[55] Nevertheless, societal concern on environmental issues during the last decade indicates that a rollback to nonenforcement is not very probable,[56] short of a profound economic crisis and depression.[57]

It would seem desirable to think of an alternative scenario, one calling for official recognition that risk is an unavoidable element of life and the common welfare, that all human lives cannot be preserved at all costs, and that carcinogenicity tests in animals cannot be reliable quantitative models of human risk. Essential elements of such an approach have been identified and debated in a recent report of the National Academy of Sciences.[58]

Today certainty in regulation is elusive, and it is

likely to remain so until adequate science develops. At the
same time, judgment in the face of uncertainty does not call
for an apology. Indeed, the current regulatory process may
be in disfavor because it is not honestly judgmental and, by
insisting on inadequate science and intransigent ideals, produces results that are perceived at times as arbitrary, inconsistent, or unacceptable to the public at large. The
central point of procedural reform is that resolution of uncertainties must be attempted in an open sociopolitical context, because usefulness, benefit, tolerable hazard, and
safety cannot be defined on the independent authority of
scientific facts or statutory prerogative.

The diversity of real-life situations would seem to make
this task impossible, but similar problems have been reconciled, traditionally, by flexible statutes that offer standards of reference, to be used in the fair and consistent
resolution of individual situations.

For the regulatory process of our interest, two sets of
references need to be defined: a standard of usefulness or
benefit and a standard of safety or tolerable hazard. After
this initial work, individual cases would be heard in open
proceedings, much as in a judiciary process.

Initially, emphasis would be on identifying functional
classes of products and uses considered necessary to sustain
a modern society. Analysis and definition of a standard of
need would have to be extensive only for each class to use.
For a particular agent it would have to be proven only that
it belongs to the class, and its standing would improve if
it offered corollary benefits, such as additional therapeu-
tic or nutritive properties. In other words, the analysis of
benefits for individual agents could be largely settled by
precedent.

The initial effort would eventually define standard
categories of use, each being assigned a relative rank of
usefulness. Primary items of need, such as basic foods,
comforts, drugs, and fuels, and perhaps basic raw materials
and chemical intermediates, would be ranked at the top,
less-needed items receiving lesser ranking, depending on a
sociopolitical judgment.

This task need not have prohibitive dimensions. It has
ample precedents in the legislative process and would appear
to be a natural function for Congress, perhaps assisted by a
systematic polling of public and expert opinion during the
extensive activities necessary at the beginning, and for
revisions thereafter.

Definitions and ranking would have to consider logistic, economic, hedonistic, esthetic, ethical, and other cultural issues. In principle these criteria would have at least equal weight in a final judgment. That man does not live by bread alone has never been so clear as in our time. The cultural mosaic of values that define happiness ought to be an important element in the definition of usefulness and benefits, even while we take into account the necessity of making choices among our desires.

Nevertheless, safety must remain an important objective, and in the process of ranking relative usefulness one could also identify a safety-standard agent representative of each class and prescribe appropriate use restrictions or tolerable conditions of exposure. This could be based on the minimum human intake, or environmental load, compatible with the fulfillment of that use, and would also depend on the rank of usefulness of the particular category. The standard-of-need agent for each class would naturally become its standard of safety or tolerable hazard, and new agents aspiring to be classified for the same use would be compared for safety with that standard. For example, sucrose could be selected as the reference for sweeteners, because it is the most widely used substance for sweetening, and because a long record of chronic exposure in mankind suggests side effects that are either ignored or accepted as tolerable by the vast majority of users. The safety of another sweetener would then be compared with that of sucrose, at doses also including maximum conditions of exposure under intended human use.

Much research would still be necessary to improve our ability to predict the relative toxic potency of two different compounds in man, because the quantitative difficulties in translating results of chronic tests across species would persist. In fact, even the direct human evidence of epidemiologic studies is not always sufficient for a regulatory decision, because of apparent or suspected confounders.

Undoubtedly, complex assay protocols would be suggested: different routes of administration in different species, extensive dose-response kinetic studies, metabolic fate determinations, structure-activity inferences, chronic and acute toxicity tests, in vitro assays with human and animal tissues, and other approaches. The redundancy of these suggestions underlines their relative impotence, besides being incompatible with the limited testing resources now available.

All things considered, it would seem reasonable that until better methods for the definition of relative toxicity

can be found, the role of science in regulation should be
limited to those instances where nearly certain assessment
of human risk is feasible and legitimate; at the same time
more emphasis should be given to methodological and basic
research for future application.

In this light, whereas carcinogenicity may not be measured reliably today, relative safety could be defined by a
formula that would assign nearly equal weights to other
forms of acute and chronic toxicity, the tests being selected
when their generalization to real life is reasonable. The
burden of proof would be left with the applicant, who could
present the case to a jury of experts and users acting in a
setting similar to a judicial proceeding to arrive at an
opinion about the toxic potency of the substance in question
relative to that of a reference agent. For the sweetener of
our example, this judiciary proceeding might succeed in defining its rank relative to sucrose, based on its relative
toxicity and the estimated dose from exposure under the intended conditons of use; and it could achieve the same rank
as the reference if, for example, it were twice as toxic but
its intended use resulted in only half the exposure.

Indeed, a safety judgment that is influenced by criteria
of need would have to consider exposure as a prime determi-
nant of hazard, and it may become necessary to develop more
sophisticated approaches for determining human intake by
various routes, under real-life conditions of an agent's
proposed use.[14]

The sum of regulatory restrictions would finally depend
on the rank of need for the class of use, more necessary ones
commanding fewer restrictions, and on the safety ranking of
the agent considered, relative to the reference compound and
the use restrictions applied to it; there might be a range
of restrictions for special situations of exposure, such as
pregnancy, young and old age, allergies, workplaces. Because
of the uncertainties, precise numerical structures for
reaching regulatory pronouncements are unlikely, even though
the formulation of decision frameworks has been discussed
and appears feasible.[58]

But who shall make regulatory decisions? This question
becomes important because the new scenario implies a shift
from normative bureaucracy to an exercise in sociopolitical
judgment. Society has repeatedly faced the challenge of
regulators preoccupied with their own survival; and traditional normative mandates have come to be questioned as rem-
nants of an autocratic past, particularly when situations
are not clear-cut but defined by a range of judgment. The

present regulatory system itself cannot avoid this situation, and in fact most of the important regulatory decisions are finally resolved by litigation.

It has been suggested[59] that it may become expedient to provide for an impartial and fair resolution of the uncertainties involved by instituting special courts independent of the regulatory agencies, the latter being left with the task of proposing regulation and enforcing the courts' decisions by devising clear categories of restriction, easily understood and accepted by all consumers and special users alike.[58] In this context, the ethical and operational incompetence of intransigent statutes might come to be viewed as an embarrassment to be rectified, and as inconsistent with the safeguards of due process that are at the philosophical core of a free society.

The proposed approach might also have important economic consequences, because of the attribution of rank to each compound within a class, and of use restrictions depending on rank. Clearly, a regulatory process that ranks efficacy and relative safety risks new dangers of intransigent interpretation. But when such a policy is exercised as social judgment, it could add a new incentive to develop increasingly better products.[58]

The future of manufacturing may well be characterized by restraints and solutions unthinkable 10 years ago and only barely felt today, chiefly reflecting the inevitable depletion of raw commodities. In that context, a new regulatory posture of the general nature suggested becomes even more plausible, as it would provide incentives for a more farsighted utilization of diminishing resources while helping to preserve the values of enterprise.

References and Notes

1. P. Pott, *Cancer Scroti: The Chirurgical Works of Percival Pott* (Clark & Collins, London, 1975), p. 734.

2. *Statistical Abstracts of the United States* (Government Printing Office, Washington, 1959), table 3; *ibid.* (1960), table 71; *ibid.* (1977), table 104.

3. R. Montesano and L. Tomatis, "Legislation Concerning Chemical Carcinogens in Several Industrialized Countries," *Cancer Research,* 37 (1977), 310-316.

4. R. Doll, "The Pattern of Disease in a Post-Infection Era: National Trends," *Proceedings, Royal Society of London,* Series B, 205 (1979), 47.

5. S. Epstein, Question and Answer Session in *Origins of Human Cancer*, ed. H.H. Hiatt, J.D. Watson, and J.A. Winsten (Cold Spring Harbor Laboratory, Cold Spring Harbor, New York, 1977), pp. 1727-1728.

6. Joint FAO/WHO Expert Committee on Food Additives, *Procedures for the Testing of Intentional Food Additives to Establish Their Safety for Use* (WHO Technical Report Series, 144; World Health Organization, Geneva, 1958).

7. "Drinking Water and Health: Recommendations of the National Academy of Sciences," *Federal Register*, 42 (1977), 35764-35779.

8. U.S., National Cancer Advisory Board, Subcommittee on Environmental Carcinogenesis, *Criteria for Assessing the Evidence for Carcinogenicity of Chemical Substances: Report of ...* (National Cancer Institute, Bethesda, MD., 1976).

9. Even current advances in mechanistic hypotheses, such as the experimentally justified multistage carcinogenesis theory, although they give useful insight into this matter, do not clarify the final molecular complexities of transformation (Refs. 12,19).

10. V. Riley, "Mouse Mammary Tumors: Alteration of Incidence as Apparent Function of Stress," *Science*, 189 (1975), 465-467; see also V. Riley and D. Spackman "Melanoma Enhancement by Viral-Induced Stress," in *The Pigment Cell*, vol. 2: *Melanomas: Basic Properties and Clinical Behavior*, ed. V. Riley (Karger, Basel, 1976), pp. 163-173; V. Riley, "Housing Stress," *Laboratory Animals*, 6 (1977), 16-21.

11. World Health Organization, *Assessment of the Carcinogenicity and Mutagenicity of Chemicals* (WHO Technical Report Series, 546; World Health Organization, Geneva, 1974).

12. J. Cairns, "Some Thoughts About Cancer in Lieu of a Summary," in *Origins of Human Cancer*, ed. H.H. Hiatt, J.D. Watson and J.A. Winsten (Cold Spring Harbor Laboratory, Cold Spring Harbor, New York, 1977), pp. 1813-1820.

13. It is in this context that precursors and proximate or ultimate carcinogens are defined as well as initiators and promotors, acting genetically or epigenetically. It appears that cellular damage leading to higher cancer frequency does not necessarily have to be of direct mutagenic nature (Refs. 18,19). As such, the effec-

tiveness of functional carcinogens could be modulated at any stage by other modifying factors which, although not carcinogenic per se, would so appear to an experimenter who notices only the resulting cancer incidence.

14. G.B. Gori, "Ranking of Environmental Contaminants for Bioassay Priority," in *Air Pollution and Cancer in Man,* ed. U. Mohr, D. Schmähl, and L. Tomatis (IARC Scientific Publications, 16; International Agency for Research and Cancer, Lyon, 1977), pp. 99-111.

15. *Guidelines for Carcinogen Bioassay in Small Rodents* (Carcinogenesis Program, National Cancer Institute, Bethesda, Md., 1976); "Scientific Bases for Identification of Potential Carcinogens and Estimation of Risk," *Federal Register,* 44 (1979), 39858-39879.

16. National Research Council, Committee for the Revision of NAS Publication 1138, *Principles and Procedures for Evaluating the Toxicity of Household Substances* (Committee on Toxicology, National Academy of Sciences, Washington, 1977).

17. Joint FAO/WHO Expert Committee on Food Additives, *Evaluation of the Carcinogenic Hazards of Food Additives: Fifth Report...*(FAO Nutrition Meetings Report Series, 29; FAO, Rome, 1961).

18. J. Cairns, *Cancer: Science and Society* (Freeman, San Francisco, 1978); R. Peto, "Detection of Risk of Cancer to Man," *Proceedings of the Royal Society of London,* Series B, 205 (1979), 111.

19. R. Peto, "Epidemiology, Multistage Models and Short-Term Mutagenicity Tests," in *Origins of Human Cancer,* ed. H.H. Hiatt, J.D. Watson, and J.A. Winsten (Cold Spring Harbor Laboratory, Cold Spring Harbor, New York, 1977), 1403-1428.

20. World Health Organization, *Procedures for Investigating International and Unintentional Food Additives* (WHO Technical Report Series, 348; World Health Organization, Geneva, 1967).

21. National Academy of Sciences/National Research Council, Drug Research Board, "Application of Metabolic Data to the Evaluation of Drugs," *Clinical Pharmacology and Therapeutics,* 5 (1969), 607-635.

22. World Health Organization, *Principles for the Testing and Evaluation of Drugs for Carcinogenicity* (WHO Tech-

nical Report Series, 426; World Health Organization, Geneva, 1969).

23. U.S. Food and Drug Administration, Advisory Committee on Protocols for Safety Evaluation, "Panel on Carcinogenesis Report on Cancer Testing in Safety Evaluation of Food Additives and Pesticides," *Toxicology and Applied Pharmacology*, 20 (1971), 419-438.

24. Ottawa, Ontario, Health and Welfare Dept., *The Testing of Chemicals for Carcinogenicity, Mutagenicity and Teratogenicity* (Ottawa, 1973).

25. C. Maltoni and G. Lefemine, "Carcinogenicity Bioassays of Vinyl Chloride: Research Plan and Early Results, *Environmental Research*, 7 (1974), 387.

26. H.F. Smyth, "Sufficient Challenge," *Food and Cosmetics Toxicology*, 5 (1967), 51-58.

27. *Appraisal of the Safety of Chemicals in Foods, Drugs and Cosmetics* (Association of Food and Drug Officials of the United States, Baltimore, 1975), pp. 79-82.

28. A. Tannenbaum, "Nutrition and Cancer," in *Physiopathology of Cancer*, ed. F. Homburger (2d ed.; Hoeber, New York 1959), pp. 517-562.

29. H.F. Kraybill, "From Mice to Men: Predictability of Observations in Experimental Systems (Their Significance to Man)," in *Human Epidemiology and Laboratory Correlations in Chemical Carcinogenesis*, ed. F. Coulston and P.E. Shubik (Ablex, Norwood, N.J., 1980).

30. E.J.C. Roe and M.J. Tucker, "Recent Developments in the Design of Carcinogenicity Tests on Laboratory Animals," *Proceedings of the European Society for Drug Toxicity*, 15 (1974), 171-177.

31. H.F. Kraybill, "Carcinogenesis Associated with Food," *Clinical Pharmacology and Therapeutics*, 4 (1963), 73-87.

32. Some official guidelines for the care of experimental animals actually prescribe lighting conditions that are known to produce rapid blindness in rats commonly used in carcinogenesis experiments, with likely profound effects on the complex neurochemical functions of the pineal gland. U.S. Dept. of Health, Education, and Welfare, *Guide for the Care and Use of Laboratory Animals* (DHEW Publication No. 7423; Washington, 1974); W.K. Noell, V.S. Walker, B.S. Kang, S. Berman, "Retinal

Damage by Lights in Rats," *Investigative Ophthamology,* 5 (1966), 450-473; W.K. Noell and R. Albrecht, "Irreversible Effects of Visible Light on the Retina: Role of Vitamin A," *Science,* 172 (1971), 76-79; W.K. O'Steen, K.V. Anderson, and C.R. Sheer, "Photoreceptor Degeneration of Albino Rats: Dependency on Age," *Investigative Opthamology,* 13 (1974), 337-339; K.V. Anderson, F. Coyle, and W.K. O'Steen, "Retinal Degeneration Produced by Low-Intensity Colored Light," *Environmental Neurology,* 35 (1972), 233-238; J. Axelrod, "The Pineal Gland: A Neurochemical Transducer," *Science,* 184 (1974), 1341-1348.

33. V. Riley et al., letter in *Science,* 200 (1978), 124-126.

34. G.E. Moore, letter in *Science,* 199 (1978), 1157.

35. A.C. Kolbye, "Cancer in Humans: Exposure and Responses in a Real World," *Oncology,* 33 (1976), 90-100.

36. The other often-advanced proposition that known human carcinogens, with notable exceptions such as arsenic, are also carcinogenic in animals, and therefore the reverse must also be true, is a generalization based on summarily matching less than two dozen recognized human carcinogens against some 2000 animal ones.

37. E.I. Goldenthal, *Current Views on Safety Evaluation of Drugs* (FDA Paper; Rockville, MD., May, 1968), pp. 1-8.

38. Food and Agriculture Organization of the United Nations, Working Party of Experts on Pesticide Residues, *Pesticide Residues in Foods* (WHO Technical Report Series, 545); World Health Organization, Geneva, 1974).

39. Federal Food, Drug and Cosmetic Act, as amended October 1976, Section 409 (C) 3A (Government Printing Office, Washington, 1976).

40. Saccharin Study and Labelling Act, Public Law 95-203, 23 November 1967.

41. G.B. Gori, "Low-Risk Cigarettes: A Prescription," *Science,* 194 (1976), 1243-1246.

42. G.E. Hutchinson, "The Influence of the Environment," *Proceedings of the National Academy of Sciences* (U.S.), 51 (1964), 930-941; B.D. Dinman, "'Non-Concept' of 'No-Threshold': Chemicals in the Environment," *Science,* 175 (1972), 495-497; G. Claus, I. Krisko, and K. Bolan-

der, "Chemical Carcinogens in the Environment and in
the Human Diet: Can a Threshold be Established?,"
Food and Cosmetics Toxicology, 12 (1975), 737.

43. Arsenic, for instance, otherwise a recognized human
carcinogen, must have a nontoxic threshold because it
is essential to the hematopoietic process and as a cat-
alyst in phosphorylation. Similar considerations are
valid for nickel, chromium, selenium, and other agents
(Ref. 26); and the presence of practical no-effect thresholds is clearly documented in smokers, not all of whom
develop lung cancer or other smoking-dependent diseases (Ref. 41). Organotropism of certain carcinogens
also implies cellular and tissue thresholds; and the
resistance of certain species to known carcinogens,
even at maximum tolerated doses, ought to be taken as
evidence that no-effect thresholds are a most common
class of real-life phenomena. Moreover, because the
carcinogenesis outcome is time- and dose-dependent,
the finite lifespans of man and animals are bound to
impose no-effect thresholds at some level of exposure
[See H. Druckrey, "Quantitative Aspects in Chemical
Carcinogenesis," in *Potential Carcinogenic Hazards
from Drugs: Evaluation of Risk,* ed. R. Truhaut (Sprin-
ger, Berlin, 1967), vol. 7, p. 60].

44. M.A. Schneiderman, N. Mantel, and C.C. Brown, "From
Mouse to Man, or How to Get from the Laboratory to
Park Avenue and 59th Street," *Annals of the N.Y. Acad-
emy of Sciences,* 246 (1975), 237.

45. D. Hoel and N. Chand, "A Comparison of Models for Determining Safe Levels of Environmental Agents," in *Reliability and Biometry,* ed. F. Proschan and R.J. Serfling
(SIAM, Philadelphia, 1974), p. 382. J. Cornfield,
letter in *Science,* 202 (1978), 1107-1108.

46. L.J. Carter, "How to Assess Cancer Risks," *Science,*
204 (1979), 811.

47. J. Walsh, "EPA and Toxic Substances Law: Dealing with
Uncertainty," *Science,* 202 (1978), 598-602.

48. C. Comar, "Bad Science and Social Penalties," *Science,*
200 (1978), 1225.

49. M.G. Morgan, "Bad Science and Good Policy Analysis,"
Science, 201 (1978), 971.

50. R.J. Smith, "NCI Bioassays Yield a Trail of Blunders,"
Science, 204 (1979), 1287-1292.

51. These tests are not yet fully developed; it is conceivable that they could yield useful data, particularly if they made use of normal human cells, tissues, and organs. Coupled with metabolic studies in man or human tissues, they might provide quantitative data more reliable for certain human extrapolations than those supplied by current animal testing (Ref. 18).

52. Toxic Substances Control Act, 11 October 1976, P.L. 94-469 (Government Printing Office, Washington, 1976); "General Provisions and Inventory Reporting Requirements, Supplemental Notice," *Federal Register*, 42 (1977), 19298; "General Provisions and Inventory Requirements," *ibid.*, p. 39182; "Notification of Substantial Risk Under Section 8(e)," *ibid.*, p. 45363; "Supplemental Notice to Proposed Inventory Reporting Requirement; Draft Reporting Forms," *ibid.*, p. 53804; "Inventory Reporting Requirements," *ibid.*, p. 64572; Office of Management and Budget, *Standard Industrial Classification Manual* (Government Printing Office, Washington, 1972); U.S. Environmental Protection Agency, Office of Toxic Substances, *Candidate List of Chemical Substances* (1977), vols. 1-3; *Addenda 1 and 2* (1978); *Activities of Federal Agencies Concerning Selected High Volume Chemicals* (1975).

53. L. Carter, "SO_2 Emissions Proposals Pose Growth Issue," *Science*, 202 (1978), 30.

54. P.H. Abelson, "The Federal Regulatory Machine," *Science*, 200 (1978), 487; P.H. Abelson, "Regulation of the Chemical Industry," *Science*, 202 (1978), 473.

55. B.J. Culliton, "Toxic Substances Regulation: How Well are Laws Being Implemented?," *Science*, 201 (1978), 1198-1199; R.J. Smith, "Toxic Substances: EPA and OSHA are Reluctant Regulators," *Science*, 203 (1979), 28-31.

56. E. Marshall, "EPA Smog Standard Attacked by Industry, Science Advisers," *Science*, 202 (1978), 949-950.

57. D.H. Meadows et al., *Limits to Growth* (Potomac Associates, Washington, 1974).

58. National Research Council, Committee for a Study on Saccharin and Food Safety Policy, *Food Safety Policy: Scientific and Societal Considerations* (National Academy of Sciences, Washington, 1979), pt. 2.

59. D.L. Bazelon, "Risk and Responsibility," *Science*, 205 (1979), 277-280.

Defining Tolerable Risk Levels

Christoph Hohenemser, Jeanne X. Kasperson

12. Overview: Towards Determining Acceptable Risk

Beyond defining and measuring risk, the issue of acceptability is the most fundamental problem for society. In
effect the question is: how safe is safe enough? Or, given
that a risk is physically and socially characterized, is it
tolerable? Defining what level is tolerable requires incorporation of human values and constraints on finite resources
in a context of uncertain information. Because tolerance
of <u>some</u> risk is unavoidable, defining acceptable risk involves placing, either explicitly or implicitly, a finite
value on life.

This part of the volume comprises four papers that
explore the problem of acceptable risk. In many respects
these are "second generation" contributions to a literature
that has been developing at least since Starr's 1969 proposal for risk/benefit analysis.[1] Each contribution, in
its own way, recognizes the failures of early optimism--
that the problem is easily solved--and hence approaches the
problem in light of contributions that have been made to
the field in the last decade.

<u>Approaches to Determining
Acceptable Risk</u>

To provide the context for the four papers in this part,
we review in the next few pages the several approaches to
determining acceptable risk that have emerged.

<u>Implicit Methods</u>

Technological risks may be accepted through processes
that involve no formal analysis at all. In this sense the
determination of risk is implicit. Examples are distribution of goods through the marketplace, the establishment of

Table 12.1. Risks that increase chance of death by one in
million.

Smoking 1.4 cigarettes	Cancer, heart disease
Drinking 1/2 liter of wine	Cirrhosis of the liver
Spending 1 hour in a coal mine	Black lung disease
Spending 3 hours in a coal mine	Accident
Living 2 days in New York or Boston	Air pollution
Travelling 6 minutes by canoe	Accident
Travelling 10 miles by bicycle	Accident
Travelling 50 miles by car	Accident
Flying 1000 miles by jet	Accident
Flying 6000 miles by jet	Cancer caused by cosmic radiation
Living 2 months in Denver on vacation from N.Y.	Cancer caused by cosmic radiation
Living 2 months in average stone or brick building	Cancer caused by natural radioactivity
One chest x-ray taken in a good hospital	Cancer caused by radiation
Living 2 months with a cigarette smoker	Cancer, heart disease
Eating 40 tablespoons of peanut butter	Liver cancer caused by aflatoxin B
Drinking Miami drinking water for 1 year	Cancer caused by chloroform
Drinking 30 12 oz. cans of diet soda	Cancer caused by saccharin
Living 5 years at site boundary of a typical nuclear power plant in the open	Cancer caused by radiation
Drinking 1000 24 oz. soft drinks from recently banned plastic bottles	Cancer from acrylonitrile monomer
Living 20 years near PVC plant	Cancer caused by vinyl chloride (1976 standard)
Living 150 years within 20 miles of a nuclear power plant	Cancer caused by radiation
Eating 100 charcoal broiled steaks	Cancer from benzopyrene
Risk of accident by living within 5 miles of a nuclear reactor for 50 years	Cancer caused by radiation

Source: R. Wilson.[2] Reprinted by permission (Copyright 1972
by the Alumni Association of the Massachusetts Institute of
Technology).

codes and licensing by local governments, the negotiation of working conditions by unions, and above all, the individual choices people make in everyday life. Taken together, these procedures have enjoyed a long history, and if the analysis of mortality time trends in Chapter 9 is any guide, they have had a major impact. Indeed, as Coates suggests in Chapter 2, despite rapid growth of formal analysis, implicit methods still predominate in determining most acceptable risk levels.

Formal Methods

The principal motivation for seeking more formal methods is that implicit methods suffer a number of defects: (1) They are not easily subjected to scrutiny; (2) they may involve serious inequity; and (3) they are poorly adapted to handling complex technological systems or serious information gaps.

Risk comparisons. A first step beyond implicit methods of determining the levels of acceptable risk involves the comparison of risks to other risks. Even though one can envision a broad definition of the concept of "risk" (see below), most risk comparisons applied to date express risk in terms of an expected annual fatality rate. A frequently cited compilation by Wilson (Table 12.1) illustrates this approach. Such risk comparisons presume that quantitative estimates are available and that agreement can be reached on the correctness of each estimate. This is not a trivial issue, since tabulations such as Table 12.1 involve a mix of experienced frequencies, correlation-derived "risk factors," and model-dependent theoretical estimates.

An obvious and attractive use of Table 12.1 is to define a threshold of acceptable risk. This may be done by noting which common risks are already accepted via implicit processes and dealing with uncommon technological risks accordingly. One may argue, for example, as did the authors of the Reactor Safety Study[3] that the background of natural disasters defines a reasonable upper bound of acceptable catastrophic risk. Nuclear reactor risks calculated in the study are then, by definition, acceptable. A difficulty with this simple argument is its failure to account for the widely held public view that reactor risks in fact exceed risks of nearly all other technologies (see Chapter 10). Another difficulty with simple risk comparison is that it fails to account for related benefits, which may be large or small. One should expect that for higher levels of benefits, higher levels of risk are tolerable.

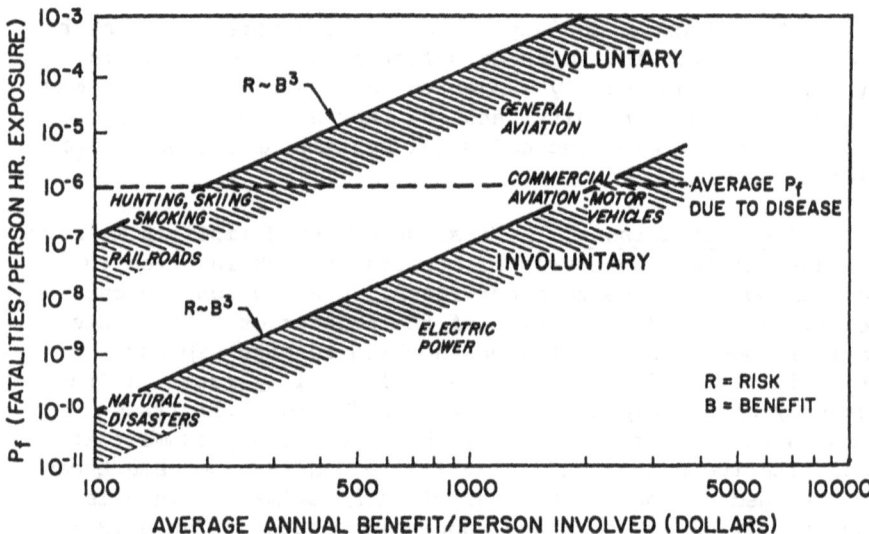

Figure 12.1. Risk versus benefit, voluntary and involuntary exposure. Reprinted by permission from C. Starr, <u>Perspectives on Benefit-Risk Decision Making</u> (National Academy of Engineering, 1972), p. 33.

Revealed preferences. To accommodate benefits, Starr[1] proposed to define acceptable risk according to the level of associated benefits. He plotted risks vs. economic benefits expressed in dollars and argued that the results "revealed" an acceptable risk standard by which other, less well-known cases could be judged. The underlying assumption is that the invisible hand of the market determines the optimal relation between risks and benefits. Starr found, as expected, that the risk, R, increased with corresponding benefit, B, as $R \sim B^3$. He also noted substantial differences between voluntary and involuntary activities and technologies. Starr's risk/benefit graph, reproduced in Figure 12.1, makes reasonable why automobile use, a voluntary risk with large benefits, has a higher level of risk acceptability than electric power, an involuntary risk with comparable benefits.

Since Starr's early work, a variety of criticisms have been made of the "revealed preference" method of determining acceptable risk. There may be common risks that are very high, yet have few associated benefits, and in this sense do not fit the picture. A plausible example of such a case is cigarette smoking. There may also be risks which through their catastrophic potential, lead to "wrong" acceptable risk levels via Starr's method. An example of this might be nuclear power. Perhaps most important, Starr's calibration of acceptable risk via common, marketplace-determined risks enshrines the old and common risks as standards, and may fail to capture a variety of aspects of new and poorly known risks that we may wish to judge by a different standard.

Risk/benefit analysis. The relation of risks and benefits suggests that an appropriate way of establishing acceptable risk is to extend the economic paradigm of cost/benefit analysis to what has become known as risk/benefit analysis. In this method both the benefits and risks are quantified in common units such as dollars, and the question is asked: do the benefits exceed the risks? Such analysis led for example to the conclusion that the risks of periodic chest x-rays exceeded the benefits: i.e., the number of TB cases detected was smaller than the probable number of cancer cases induced.

Risk/benefit analysis is attractive because it makes explicit some assumptions that are normally hidden. Risk/benefit analysis is problematical, because except for favorable cases such as the TB-cancer trade-off cited, it requires placing a dollar value on human life in order to allow direct comparison of "lives saved" to "benefits received." Risk/benefit analysis breaks down altogether when risks or benefits cannot be reliably quantified, as in the case of toxic chemicals. Risk/benefit analysis can lead to useless results when distribution of risks and benefits is significantly skewed, so that one population segment bears most of the risks while another receives most of the benefits. Most fundamentally, risk/benefit analysis fails when the concept of "expected fatality rate" does not adequately capture the full complexity of the idea of risk, such as is obviously the case for nuclear technology.

Analysis of cost effectiveness. To avoid the full complexity of risk/benefit analysis, but nevertheless to transcend the logic of simple risk comparisons, it is possible to envision intermediary methods. One of these, known as cost-effectiveness ranking, concentrates on defining regulatory options rather than the level of acceptable risk itself. Control actions are arrayed in the order of risk reduction achieved per dollar spent. In effect, the method does not deal adequately with the social dimensions of risk and may lead to a policy that violates public preferences (see below).

Aversion. An extreme case of risk/benefit analysis occurs when an activity or technology has only risks and no benefits or comes close to that state. Here it makes sense to scrap all formal analysis and prohibit the activity or technology without further argument. In the view of many people, handguns are a case in point. In the view of the authors of the Delaney amendment, carcinogenic food additives are another example. Yet, as has been illustrated in both of these cases, and in others where banning has been attempted, the argument for prohibition is never so tight as to leave no room for controversy and hence ultimate rejection of the policy. Thus we have discovered that some carcinogens are also of significant benefit and that many individuals, in fact, see remarkable advantages to unlicensed use of handguns.

Methods Incorporating Public Preferences

If there is a single criticism that may be made of the methods for determining acceptable risk, it is that risk has been too narrowly defined. As already noted in the introduction to this volume, the definition of risk that an analyst adopts is not a matter of semantics but leads directly to what will be studied and measured. To date the analysis of acceptable risk has been dominated by people who define risk narrowly as a frequency or probability of death or injury.

Expressed Preferences. A number of social scientists have shown that in the subjective judgment of ordinary people, risk is a multidimensional concept that cannot be meaningfully reduced to "expected mortality" (see Chapter 10). Out of this insight has grown the realization that a proper analysis of acceptable risk involves the question: "How safe is safe enough according to the expressed preferences of people?" Though lay preferences determine risk levels in many cases where individual choice is dominant (driving, drinking, smoking), "expressed preferences" is most significant as a check on the formal methods we have mentioned. Fischhoff and collaborators in a classic paper,[4] and several other workers since then,[5] have shown that expressed preference studies may lead to results that are significantly at variance with formal methods based on a narrow definition of risk as probability. These results can be understood in terms of a variety of risk characteristics, such as the degree of familiarity, the number of people potentially killed in a single event, whether future generations are affected or not, and other similar attributes. The important point to recognize when contemplating the results of expressed preference studies is not that lay people are irrational, but that they employ a definition of risk that is broader than that of most risk analysts.

Adversarial process. The question of acceptable risk is increasingly being settled through adversarial processes, often in the courts. This applies, for example, to several cases of nuclear power plant siting, to the transport and disposal of radioactive wastes, to the handling of toxic wastes, and most recently to the appropriateness of a workplace standard set by a federal regulatory agency. Adversarial processes go beyond implicit methods in that risk levels are often explicitly debated; and they go beyond individual formal methods in that the results of several kinds of analysis, including some form of expressed preferences, are likely to be incorporated. The main disadvantage of adversarial proceedings is that they are very costly and time-consuming and cannot be applied to most routine cases of acceptable risk determination. One may expect, however, that they will continue to dominate in the future for cases which are (a) too important to be left to the experts, (b) too complex to yield clear answers via formal analysis, and (c) strongly affected by public preferences that run counter to the results of other analytic methods.

Contributions on Risk Acceptance
================================

The papers in this part include two contributions on
acceptable risk generally, one on the problem of technologically related cancer risk, and one on the issue of cost-
effectiveness ranking. All four papers define "risk" narrowly as an expected fatality rate, although some authors
note that there are broader, social dimensions of risk
which may also be relevant.

    Leading off the series, Okrent points out that many
risks are poorly known and that this hampers decision making. He believes that better risk quantification will elicit
better decisions. To illustrate, Okrent cites from his own
experience a number of catastrophic risks, including a petrochemical complex in Britain, dams and earthquakes in California, liquid natural gas storage for Los Angeles, and
flood control in the Front Range canyons of Colorado. To
define the level of risk that is acceptable, Okrent proposes a regulatory procedure that incorporates quantitative
risk standards. Although Okrent recognizes that risk acceptance is a complex problem that might well entail a broad
range of considerations, his own proposal is essentially a
risk comparison based on estimated fatality rates. As such
it is a throwback to a time before risk/benefit analysis
was invented, and might be seen as an expression of frustration with risk/benefit procedures. As Okrent argues co-
gently, these have resulted in "few specific proposals for
workable, defensible, risk/acceptance criteria."

    In the following paper, Starr and Whipple explore the
limits of risk/benefit decision-making. The authors detect
in risk management an overall failure to mesh informal or
"intuitive" risk/benefit processes with analytic approaches
such as formal risk/benefit analysis. As a measure of
these intuitive processes they cite work (represented in
Chapter 10 of this volume) on risk perception by Slovic,
Fischhoff, and Lichtenstein, as well as related studies on
"expressed preferences" that compare perceived risk and
perceived benefit.[4] To circumvent the mismatch between
expressed preferences and analytic risk/benefit techniques,
Starr and Whipple, like Okrent, propose quantitative risk
acceptance criteria below which regulatory intervention
would not be required. They regard quantitative criteria
as attractive

    because the key to acceptability of a technology
    under the proposed method is the level of risk.
    Assuming that the estimated risk became the central

point in the debate, the public might have more
confidence in the regulatory systems if their concern is directly addressed.

In effect, Starr and Whipple recognize that rational model I
(risk/benefit analysis) is unworkable in cases of conflict
between analytic and expressed preference methods, and seek
to substitute rational model II (quantitative risk criteria)
for it.

One can appreciate the neat simplicity of Okrent-Starr-
Whipple proposals. One can, however, see the difficulties.
Setting quantitative criteria assumes that appropriate probabilities can be derived or calculated. It also assumes
that expected fatality rates will be accepted by the public
as appropriate measures of risk. Though neither Okrent nor
Starr and Whipple say it, the latter requires a major shift
in public attitudes in which the complexity of judgments
either disappears or is somehow suppressed. This is not
to suggest that the call for quantitative risk standards is
naively based. For example, Starr and Whipple close their
argument by urging further understanding of expressed preference risk/benefit judgments and the use of such understanding to rationalize a double standard of risk in certain
cases like nuclear power.

In the third paper, Deisler addresses the problem of
acceptable risk for industrially related carcinogens. He
recognizes at the outset that we do not currently know the
answer to the question: how much cancer is industrially
related? His survey of the literature yields estimates--
consistent with those given by Hohenemser, Kasperson, and
Kates in Chapter 9--that range between 1 and 38% of all
cancers. To motivate his analysis, Deisler characterizes
current federal regulatory policy on carcinogens as consisting of two alternative prescriptions: (1) "set residual
risk so that there is 'no real risk' to people" (2) "lower
the risk to the lowest feasible level." He points out that
the first may involve great effort in going from 10^{-5} to 10^{-6}
in expected cancer rate per individual, at a very high cost
per life saved, and that the second leaves undefined what
is feasible and thus begs the question.

As an alternative, Deisler, as do Okrent, Starr and
Whipple, proposes quantitative risk acceptance levels. He
suggests a goal of reducing the risk from industrially
related carcinogens to 1% of the total cancer rate and
spends considerable effort in formulating how that goal
might actually be achieved. His scheme relies intimately

on availability of dose-response data including synergistic
effects for individual carcinogens. In this he recognizes
the impossibility of obtaining direct human data and assumes that animal data, however deficient, will be used.
Deisler's contribution is particularly valuable in its
delineation of the kind of information that will be required
to set quantitative risk standards for industrial carcinogens; at the same time, it includes no actual case applications.

In evaluating Deisler's proposal it is helpful to consider the commentary of Gori on cancer regulation (Chapter
11). Along with Deisler, Gori recognizes that present federal regulatory policy is based not on risk/benefit considerations, but on a quasi-absolute, near zero risk aversion rules that are inadequate to the task. Gori does not
assume, however, that carcinogenic risk can be quantita-
tively defined in most cases. Rather, he reviews the many
apparently fundamental reasons why efforts by health scientists to quantify human risk via animal experiments are
bound to fail. Attempts to implement Deisler's logical
and straightforward program may well have to counter the
dissent of a significant part of the scientific community.

In the final contribution to this part, Schwing dis-
cusses the problem of cost-effectiveness analysis, and substitutes "increased longevity" for the more frequently
used "lives saved" as a measure of success for risk management. Schwing argues that longevity is a more intuitive
concept than lives saved. Schwing's technical contribution
is an apparently simple transformation between crude mortality and increased longevity that permits estimation of
the latter if the former is known. Schwing illustrates his
method with a number of informative examples chosen from
technological risk cases for which annual mortality figures
are available. As have other cost-effectiveness studies
of risk management, Schwing's paper points up how present
mixed priorities lead to an expenditure of effort that in
cost-effectiveness terms, appears nonoptimum.

Such a finding, however, lends itself to two quite different interpretations: (1) The great variability in cost
effectiveness of management may indicate the need for management reform toward a more rational, resource-optimizing
direction. (2) Alternatively, the great variability in
cost effectiveness of management may be an artifact of the
narrow definition of risk used; for perhaps management is
already responding to the complex definition of risk preferred by the general public. The Department of Transpor-

tation, for example, currently spends enormous amounts on widening highways and almost nothing on increasing seatbelt usage, even though the latter is far more cost-effective.[6] Is this "irrational," or merely in accord with public preference?

If there is one word that summarizes the general question of determining risk acceptability, it is "unsolved." For, though our contributors reflect many of the current concerns of risk analysts, there is as yet no adequate synthesis that leads to an integration of the many ways of looking at the problem. Most in need of further clarification, in our view, is the relation between expressed preferences and formal methods. Here what really seems to be at stake is the very definition of "risk." If risks are threats to people and what they value, should they be measured by statistical data defining expected mortality or morbidity? Or should we use a broader concept that incorporates a number of other qualities apparently considered relevant by the general public? To put it differently: Should the general public be educated to accept the simple beauty of Wilson's table (Table 12.1), or should risk analysts be educated to accept the multivariate meaning of risk that pervades public thinking?

References and Notes

1. C. Starr, "Social Benefit versus Technological Risk," *Science,* 166 (1969), 1232-1238.

2. R. Wilson, "Analyzing the Risks of Daily Life," *Technology Review,* 81 (February, 1979), 45.

3. U.S. Nuclear Regulatory Commission, *Reactor Safety Study: An Assessment of Accident Risks in U.S. Commercial Nuclear Power Plants, Executive Summary* (WASH-1400, NUREG-75/014; Washington, October 1975).

4. B. Fischhoff, P. Slovic, S. Lichtenstein, S. Read, and B. Combs, "How Safe is Safe Enough?: A Psychometric Study of Attitudes Towards Technological Risks and Benefits, *Policy Sciences,* 8 (1978), 127-152.

5. *Societal Risk Assessment: How Safe is Safe Enough?,* ed. R.C. Schwing and W.A. Albers (Plenum Press, New York, 1980).

6. U.S. Dept. of Transportation, *The National Highway Safety Needs Report* (Washington, D.C., 1976).

David Okrent

13. Comment on Societal Risk

The terms "hazard" and "risk" can be used in various
ways. Their usage in this article is defined by the following simple example.

Three people crossing the Atlantic in a rowboat face a
hazard of drowning. The maximum societal hazard in this
case is three deaths. Three hundred people crossing the
Atlantic in an ocean liner face the same hazard of drowning,
but the maximum societal hazard is 300 deaths. The risk to
each individual per crossing is given by the probability of
the occurrence of an accident in which he or she drowns.
The risk to society is given by the size of the societal
hazard multiplied by the probability of the hazard. Clearly
the hazard is the same for each individual, but the risk is
greater for the individuals in the rowboat than in the ocean
liner.[1]

Some general observations follow:

1) Society is not risk-free and cannot be. No energy
source is free of risk, either to the environment or to the
public. This includes solar energy.[2] Measures toward
achieving "soft" energy, zero increase in energy consumption,
and even conservation inherently carry risk.

2) There are large gaps in society's understanding of
risks and the economics of risk management.[3] Risk-benefit
analysis, in the legislative process and elsewhere, is an
important tool in decision-making, and should be judiciously
employed. Procedures are needed to ensure proper disclosure
of assumptions, uncertainties, unaggregated results, and so

Reprinted by permission from Science, Vol. 208 (25 April 1980),
pp. 372-375. Copyright 1980 by the American Association for
the Advancement of Science.

forth, and to ensure impartial evaluation and review of any
important risk-benefit decisions.

3) The consequences of two different hazards may vary
greatly with respect to their measurability. This problem
(together with the possible erroneousness of the raw data),
has led to a questioning of the desirability of using such
information in making decisions. Furthermore, some hazards,
such as the greenhouse effect or the effect of energy policies on the chance of war,[4] introduce risks that can be difficult to quantify. Nevertheless, for many societal hazards,
risk quantification, albeit imperfect and frequently containing large uncertainties, is usually possible and desirable. Decisions still have to be made, and they are likely
to be better if they are made with the benefit of more complete information--keeping in mind the need for judgment
as to when the inability to quantify completely may lead to
unwarranted dependence on estimates. Consideration should
be given to requiring risk quantification, as practical, for
societal endeavors.

4) Society uses the word safe in a vague and inconsistent fashion. Efforts to reduce risk are not necessarily
made in the most cost-effective way. Our priorities should
be reevaluated.

5) In view of their statistically smaller contribution
to societal risk, major accidents may be receiving propor-
tionately too much emphasis compared to other sources of
risk, such as chemical residues, pollutants, and wastes.

6) Society's resources are limited. When resources are
lavished on a needed service, less is available for use in
measures that reduce the number of injuries and premature
deaths. Thus a more expensive source of electricity carries
an economic penalty compared to a cheaper source. Above a
particular level, expenditure of resources on additional
programs to reduce risks to health and safety may be counterproductive because of adverse economic and political effects.

7) Congress should take the lead in establishing a
national risk management program that is equitable and more
quantitative.

The Need for Information

There are few published assessments of the many hazards
and risks to which society is exposed. And there are still
fewer risk assessments that (i) provide a detailed statement
of the assumptions made in arriving at the conclusion, (ii)

point out any uncertainties in the results, and (iii) have
the benefit of a detailed evaluation by a competent independent body.

For example, it is difficult to find published quantitative estimates of the risks posed by the thousands of
large dams in the United States. In fact, the safety of
such dams is generally poorly known, particularly with respect to the more serious, lower probability modes of failure. The situation is the same for facilities in which large
amounts of hazardous chemicals are stored.

We also know little about the risks created by emissions
of substances into the atmosphere, by the disposal of liquid
and solid wastes, by coal-fueled electric power stations, by
residues and additives in our food, by occupational environ-
ments--and the list could go on.

Nevertheless, substantial improvement can be made in
our knowledge about risks and the costs of their reduction.

Examples of Hazard and Risk Estimates

Canvey Island. An interesting and significant risk
study, *Canvey: Summary of an Investigation of Potential
Hazards from Operations in the Canvey Island/Thurrock Area*,[5]
was released in June 1978 by the Health and Safety Executive
of the British government.

Canvey Island lies in the Thames River and is 9 miles
long and 2.5 miles wide. It has 33,000 residents and seven
large industrial complexes, including petroleum, ammonium
nitrate, and liquefied natural gas facilities. The largest
risk of death from an accident at one of these industrial
facilities was estimated to be about 1.3×10^{-3} (1 in 800)
per year for some of the nearest Canvey residents. This risk
is about five times as large as the average risk of dying in
an automobile accident in the United States. The average
risk of death from an accident at these installations was
estimated to be about 5×10^{-4} (1 in 2000) per year for all
the island's residents. This is about twice the risk of
death from an auto accident in the United States. The chance
of 1500 people being killed in a single accident was given
as more than 1 in 1000 per year. The chance of 18,000 being
killed in a single accident was given as 1 in 12,000 per year.

It was stated that these estimates probably erred on
the side of pessimism by a factor of 2 or 3, but probably
not by a factor of 10. The Health and Safety Executive
recommended that improvements be made that would reduce the

likelihood of each of the risk estimates by a factor of 2 or
3. With these improvements, it was judged, the risks would
be acceptable.

My discussions with British experts in safety assessment
have given me the impression that they doubt the practicality
of obtaining improvement, for every large facility in the
British chemical industry, by more than a factor of 10 over
the risks estimated for Canvey. However, the British are
making it a matter of national law that safety assessment reports be submitted by each industrial facility utilizing or
storing more than a particular quantity of a hazardous chemical. Notification is still required if some specified lesser quantity is stored. The Health and Safety Executive will
have the responsibility for evaluating the risk assessment
and deciding on the acceptability of the risk.

Japan is also instituting safety design requirements
for chemical plants, requirements that become increasingly
strict in proportion to the number of deaths that might occur if there is a serious accident.

Should not the United States be developing some systematic approach to these and other societal risks? I have
little doubt that we have many chemical installations posing
risks not unlike those at Canvey.

<u>Dams in California</u>. Limited studies of ten California
dams by our group at the University of California, Los
Angeles, indicated that up to 250,000 deaths could result
from catastrophic failure of the largest of these dams.[6]
Historically, large dams have failed (although not necessarily suddenly and in gross fashion) at a rate of about 1
in 5000 per year. However, our crude estimates of the failure rate for some of the dams studied were as large as 1 in
100 per year.

During the San Fernando Valley earthquake in 1971, the
Van Norman Dam nearly failed catastrophically due to soil
liquefaction, a phenomenon recognized only after its construction. Had the reservoir been full, the dam would have
failed,[7] possibly causing 50,000 to 100,000 fatalities.

The state of California has had a dam-safety law since
the 1971 earthquake. The law specifies that the safety of
each state-controlled dam must be reviewed and a finding of
"safe" made. However, under the law the state need not pub-
licize the risk it is imposing when it determines that a dam
is safe. And, of course, the maximum possible number of
fatalities remains unchanged by any finding.

Earthquakes in California. California faces serious
safety questions concerning the possibly catastrophic effect
of earthquakes on its cities. This is also true in other
states, but in California the problem is acute. On 17 March
1976 the U.S. Geological Survey advised Governor Brown of
the relatively large likelihood that a major earthquake in
Los Angeles would kill many thousands of people, primarily
from collapse of seismically substandard buildings and from
dam failure. A report prepared for the Federal Disaster
Assistance Administration makes equally gloomy predictions.[8]
To my knowledge, seismically substandard buildings have not
been posted as hazardous in Los Angeles, nor have instructions been issued on where to go in the event of dam failure.
The city of Los Angeles, of course, has been grappling with
the problem of seismically substandard buildings for
years.[9,10] Seismic retrofit or building condemnation is
very costly. So far as I know, the state of California
has not devoted significant financial resources to this
problem.

Liquefied natural gas (LNG). The LNG technology has
been the object of increasingly intense safety review in the
past few years. One of the few large proposed U.S. chemical
installations for which a serious, detailed risk study has
been published is the LNG facility for Los Angeles,[11]
Oxnard, or Point Conception, California.

The study, which was performed under a contract for the
corporation requesting to build the facility, has been a
subject of considerable controversy. It did not include a
self-critique in which assumptions were clearly identified
and uncertainties critically evaluated. And it did not
establish quantitative criteria against which to judge the
acceptability of the risk at each of the proposed sites.

Although other risk estimates for these proposed facilities have been issued (most project larger risks), a detailed study and evaluation by an independent group is not,
to my knowledge, available.

The state of California has imposed very stringent
siting requirements for LNG facilities, but California makes
no systematic assessment of the hazards from large chemical
installations, some of which may pose risks similar to or
greater than those from the previously proposed LNG facility in Los Angeles harbor. Thus, in a state that is relatively advanced in its efforts to control risk to the public, it is difficult to find a uniform rationale for the
standards, priorities, and resources used in this job.

The Flood at Big Thompson Canyon

Some may call the flood at Big Thompson Canyon, Colorado,[12] a natural disaster. But if most of the fatalities could have been prevented by proper advance planning or emergency action, I am unwilling to shrug off the event as a natural disaster, seemingly beyond our control and not to be compared with accidents in man-made facilities.

During the evening of 31 July 1976, an intense thunderstorm stalled over a small portion of Big Thompson Canyon, dropping ten or more inches of rain in a 3-hour period. Because of the steep mountain topography, the runoff quickly formed a virtual wall of water that displaced everything in its path. Of about 4000 people in the canyon, 139 died and 4 were never found. Property damage exceeded $41 million.

The area was totally unprepared for such an event. Efforts to evacuate were made, but they were obviously inadequate. Was the loss of life the result of a natural catastrophe that could not be avoided? It might have been avoided altogether by restrictions on building on the floodplain--a controversial matter. Accepting the de facto use of the floodplain, the loss of life could still have been minimized with the benefit of some prior analysis, a reasonably direct method of measuring and monitoring rainfall, and a suitable warning system.

I do not recall any congressional investigation of the matter. Colorado has since imposed restrictions on rebuilding in the floodplain in Big Thompson, but these restrictions are being fought. There are many similar canyons all along the Front Range of the Rockies, including one that opens onto Boulder, Colorado. What safety precautions are being taken for these canyons and for other similar "natural" hazards? Is this question being given the same priority as new LNG facilities or nuclear power plants?

Expenditures to "Save a Life"

The expenditures made by society to save a single life vary to a remarkable degree. Morlat[13] estimated that in France, $30,000 was being spent per life saved through road accident prevention and about $1 million per life saved through aviation accident prevention. Sinclair[14] estimated that in Great Britain the expenditures ranged from $10,000 for an agricultural worker to $20 million for a high-rise apartment dweller.

Comparable disparities among implicit values of life

are easily found in the United States. In a report prepared by the National Academy of Sciences (NAS) for the
Senate Committee on Public Works in March 1975, estimates
were made of the health costs of the pollutants from coal-
fired electric generating plants.[15] A figure of $30,000
per premature death was used,[15(p.611)] "rather than the value of $200,000 used in highway safety." The reasoning for
this choice was that "most of the deaths occur among chron-
ically ill, elderly people, and the amount by which their
lives are reduced may be only a matter of days or weeks."
This value of life was then used as the reference value for
cost-benefit trade-offs that provided a basis for evaluating
the merits of various approaches to control of emissions
from coal plants, including the timing of such controls.
The estimated number of premature deaths resulting from the
activities of coal plants was lower in the NAS report than
the highest estimate given in other publications.[16] The
actual value is quite uncertain.

On the other hand, in its "As low as reasonably achievable" (ALARA) criterion for routine releases of radioactivity from a nuclear power plant, the Nuclear Regulatory Commission (NRC) employs $1000 per man-rem as the expenditure
limit for making improvements. On the basis of estimates
from the BEIR report,[17] this translates into more than $5
million per premature death deferred. Furthermore, this
death would probably occur after the age of 50; hence, if
one remained consistent with the philosophy promulgated in
the NAS study, the $5 million derived from the NRC criterion
should be compared to a value less than the $200,000 quoted
in the NAS report.

The societal risk from the disposal of hazardous liquid
and solid wastes is substantial. I doubt that society is
using the same risk-acceptance criteria or value of life in
its choice of criteria for disposal of radioactive and nonradioactive wastes. I believe that a similarly large discrepancy exists with respect to regulation of the transportation of hazardous radioactive and nonradioactive materials.

Resource Allocation

Resources for the reduction of risks to the public are
not infinite. At some point, a greater improvement in
health and safety is to be expected from a more stable and
viable economy than from a reduction in pollution or the
rate of accidents. For example, Siddall[18] recently showed
a direct correlation between increased life expectancy and
improved economic circumstances in Great Britain.

Perhaps Congress should initiate appropriate studies to enable a reasonably accurate evaluation to be made of the proper level of expenditure for risk reduction. Within such a level of expenditure, if we fail to devote our resources to those risks in which the most reduction is achieved per dollar, we are not optimizing the effect of our capital outlay.[19],[20] Of course, one must ensure that there are no gross inequities; that no individual is knowingly left exposed to a risk significantly greater than some upper limit of acceptability.

Each individual or group that makes recommendations, or otherwise takes actions affecting national priorities, bears some responsibility for any adverse effects. Thus an individual who effects the banning of DDT in a tropical country may inadvertently cause far more deaths than he defers, since the incidence of malaria will then increase. Similarly, if coal-burning electric generating plants are found to cause far more premature deaths than nuclear power plants (in agreement with most published estimates), an individual or agency that successfully advocates the construction of coal-burning plants instead of nuclear power plants may be responsible for unnecessary deaths. If the media should present an unbalanced perspective on some aspect of risk in society, and this causes risk-reduction priorities to be set inefficiently and even wrongly, the responsible media would, in effect, be contributing to the causing of premature deaths that might otherwise have been averted.

Approaches to Risk Acceptance

Lowrance[3] said, "A thing is safe if its risks are judged to be acceptable."

The Van Norman Dam was presumably considered to be safe before it nearly failed in 1971. Was it safe?

The *Los Angeles Times* some years ago editorialized concerning the proposed Auburn Dam, saying, "Let's build it if it's safe." What does safe mean in the context of an Auburn Dam whose failure was estimated by an experienced engineer to be capable of killing 0.75 million people?[21] We cannot prove that there is zero probability of its failure. What estimated failure probability is acceptable? What level of uncertainty in this estimate is acceptable? Will it be possible to demonstrate that such a safety goal can be achieved?

The NRC licensed reactor No. 2 at Three Mile Island before the accident there. Hence it had determined that

"there is reasonable assurance that the activities authorized by this operating license can be conducted without en-
dangering the health and safety of the public." However,
the NRC did not provide an estimate of the residual risk remaining after the inclusion of required safety features.
And the NRC still has not qualified its definition of "reasonable assurance" with substantive numbers.

The approaches society might use in coping with "How
safe is safe enough?" included (i) nonintervention (rely on
the marketplace), (ii) professional standards (rely on the
technical experts), (iii) procedural approaches (muddle
through), (iv) comparative approaches (reveal or imply preferences), (v) cost-benefit analysis, (vi) decision analysis,
and (vii) expressed preferences (rely on public perception
of risk).

It is to be anticipated that any generally accepted approach would incorporate facets of most of the above, as appropriate. It is not proposed that quantitative risk-acceptance criteria can or should represent the whole approach.
However, they should play an important role.

It is not easy to develop a workable, defensible set of
quantitative risk-acceptance criteria that also allow for
benefits, societal needs, equity, economics, political and
social effects, and so forth. As a result, few specific
proposals have been published. In an effort to stimulate
discussion on the subject, Okrent and Whipple[22] described a
simple quantitative approach to risk management that incorporates the following principal features:

1) Societal activities are divided into major facilities
or technologies, all or part of which are categorized as essential, beneficial, or peripheral.

2) There is a decreasing level of acceptable risk to
the most exposed individual (for example, 2×10^{-4} addition-
al risk of death per year for the essential category, 2×10^{-5} for the beneficial category, and 2×10^{-6} for the peripheral category).

3) The risk is assessed at a high level of confidence
(say 90 percent), thereby providing an incentive to obtaining better data. (The expected value of risk must be smaller the larger the uncertainty.)

4) Each risk-producing entity is subjected to risk assessment in terms of both the individual and society. The
assessment is performed under the auspices of the manufac-

turer or owner but must be reviewed and evaluated independently; the decision on acceptability is made by a regulatory group. (For practical reasons, there would be some risk threshold below which no review was required.)

5) The cost of the residual risk is internalized, generally through a tax paid to the federal government, except for risks that are fully insurable and, like drowning, readily attributable.

6) The government, in turn, redistributes the risk tax as national health insurance or reduced taxes to the individual.

7) Risk aversion to large events would be built into the internalization of the cost of risk, but with a relatively modest penalty. If a technology or installation poses a very large hazard at some very low probability (and many do), case-by-case decisions are made, with considerable emphasis on the essentiality of the venture.

8) An ALARA criterion on risk is required, although an incentive to reduce risk and associated uncertainties would already be provided by establishing a suitable level for the risk tax.

This quantitative approach to risk management is, of course, untested. It may be both too complex and too simple. It is subject to the obvious difficulty of defining what constitutes a risk-producing entity. However, there has been all too little real discussion of the question, "How safe is safe enough?" Comar[23] suggested a "de minimis" approach, and there are a few other proposals. But, more typically, entire symposia are held on risk management without so much as mentioning the subject of quantitative risk criteria.

In conclusion, if our priorities in managing risk are wrong, if we are spending the available resources in a way that is not cost-effective, we are, in effect, killing people whose premature deaths could be prevented. There is some optimal level of resources that should be spent on reducing societal risk, a level beyond which adverse economic and political effects may be overriding. Finally, there is need for the development of a national approach to risk management, one that Congress, the President, and the public can support.

References and Notes

1. This example is, of course, simplified with regard to the hazards faced on such a journey. Those in the rowboat face the possibility of dehydration and starvation, while those on the ocean liner are subject to fire and falling overboard. Other hazards can be imagined.

2. In 1957 I designed my own passive solar home in northern Illinois. It was very well insulated and had the tightest storm windows available. Not till 20 years later did I become aware that I had thereby been exposing my family to increased indoor air pollution. I have learned to be a skeptic about risk. I am particularly skeptical of those who advocate a particular technology as benign, or attack a technology as too risky, without presenting a detailed, quantitative risk evaluation, without making a choice among feasible alternatives, and without placing the risks in some broader societal perspective.

3. Modern risk-benefit thinking had its real birth with the classic paper by C. Starr, "Social Benefit versus Technological Risk," *Science,* 165, (1969), 1232-1238. For background material, see *Perspectives on Benefit-Risk Decision Making* (National Academy of Engineering, Washington, D.C., 1972); W.W. Lowrance, *Of Acceptable Risk: Science and the Determination of Safety* (Kaufman, Los Altos, California, 1976); W.D. Rowe, *An Anatomy of Risk* (Wiley, New York, 1977); A.J. Van Horn and R. Wilson, *The Status of Risk-Benefit Analysis* (BNL 22282; Brookhaven National Laboratory, Upton, New York, 1976); D. Okrent, Ed., *Risk-Benefit Methodology and Application: Some Papers Presented at the Engineering Foundation Workshop* (UCLA-ENG-7598; University of California, Los Angeles, 1975); *How Safe is Safe? The Design of Policy on Drugs and Food Additives* (Academy Forum, 1; National Academy of Sciences, Washington, 1974); D. Okrent; *Final Report: A General Evaluation Approach to Risk-Benefit for Large Technological Systems and Its Application to Nuclear Power* (UCLA-ENG-7777; University of California, Los Angeles, 1977).

4. I favor a national energy approach that emphasizes conservation and the wise use of all domestically available resources. Diversity will better ensure resiliency. Too much discussion has made it appear that the United States must choose between solar and nuclear energy or

between the "soft" or "hard" energy paths. We need solar and nuclear and coal energy.

5. Great Britain, Health and Safety Executive, *Canvey: Summary of an Investigation of Potential Hazards from Operation in the Canvey Island/Thurrock Area* (Her Majesty's Stationery Office, London, 1978).

6. P. Ayyaswamy, B. Hauss, T. Hsieh, A. Moscati, T. Hicks, D. Okrent, *Estimates of the Risks Associated with Dam Failure* (UCLA-ENG-7423; University of California, Los Angeles, 1974).

7. H. Seed, H.B. Seed, K.L. Lee, I.M. Idriss, "Preliminary report on the lower San Fernando Dam slide during the San Fernando earthquake of February 9, 1971," Appendix B in *Interim Report* (California Department of Water Resources, Sacramento, 1971).

8. S.T. Algermissen, M. Hopper, K. Campbell, W.A. Rinehart, D. Perkins, *A Study of Earthquake Losses in the Los Angeles, California Area: A Report Prepared for the Federal Disaster Assistance Administration, Department of Housing and Urban Development* (Environmental Research Laboratories, National Oceanic and Atmospheric Administration, Rockville, MD, 1973).

9. K.A. Solomon, D. Okrent, M. Rubin, *Earthquake Ordinances for the City of Los Angeles, California: A Brief Case Study* (UCLA-ENG-7765; University of California, Los Angeles, 1977).

10. Earthquake disaster studies have also been prepared for Salt Lake City and the Puget Sound area. Most major U.S. cities are vulnerable to catastrophic damage from an earthquake, although the likelihood is highest in Los Angeles and San Francisco.

11. *LNG Terminal Risk Assessment Study for Los Angeles, California* (SAI-75-614-LJ; Science Applications, Inc., Los Angeles, 1975).

12. D.B. Simons et al., *Flood of 31 July 1976 in Big Thompson Canyon, Colorado* (National Academy of Sciences, Washington, 1978).

13. G. Morlat, "Un modèle pour certaines décisions médicales," (Cahiers du Séminaire d'Économetrie; Centre Nationale de la Recherche Scientifique, Paris, 1970).

14. C. Sinclair, P. Marstrand, P. Newick, *Innovation and Human Risk: The Evaluation of Human Life and Safety in Relation to Technical Change* (Centre for the Study of Industrial Innovation, London, 1972).

15. "Air quality and stationary source emission control," prepared for the U.S. Senate Committee on Public Works by the Commission on Natural Resources, National Academy of Engineering, Washington, D.C., 1975.

16. C.L. Comar, and L.A. Sagan, "Health Effects of Energy Production and Conversion," *Annual Review of Energy,* 1 (1976), 581-600.

17. National Research Council, Advisory Committee on the Biological Effects of Ionizing Radiations, *Considerations of Health Benefit-Cost Analysis for Activities Involving Ionizing Radiation Exposure and Alternatives: A Report of* ...(National Academy of Sciences, Washington, 1977).

18. E. Siddall, *Nuclear Safety in Perspective* (Canatom, Toronto, Ontario, Canada, 1979).

19. R.C. Schwing, "Longevity Benefits and Costs of Reducing Various Risks," *Technological Forecasting and Social Change,* 13 (1979), 333-345; see also Chapter 16, this volume.

20. R. Wilson, testimony at hearings of the Occupational Safety and Health Administration, Washington, D.C., April 1978.

21. H. Cedergren, testimony at the hearing of the Water Projects Review Committee, Sacramento, California, 21 March 1977.

22. D. Okrent and C. Whipple, *An Approach to Societal Risk Acceptance Criteria and Risk Management* (UCLA-ENG-7746; University of California, Los Angeles, 1977).

23. C.L. Comar, "Risk: A Pragmatic *de minimis* Approach," *Science,* 203 (1979), 319.

14. Risks of Risk Decisions

Technology creates many risks. Determining which risks are acceptable is an important national issue. It pervades major sectors of our economy: In food production we face decisions about pesticides and preservatives; transportation risks are increasingly regulated; and a central issue in energy policy is the controversy over the risks from power plants. Regardless of whether the seriousness of technological risk is only now being recognized, or, alternatively, that the preoccupation with risk and regulations is an over-reaction, it is clear that the cost to society of the conflict over accepting technological risks is great. These costs stem from the anxiety suffered by those who are dismayed by the conflicting information about these risks, and from the litigation, misplaced investment, retrofits, and costly delays that result from industry's inability to predict the acceptance of risk by the public.

Risk assessment is growing in importance as a system design tool. The final configuration of all technical systems is the outcome of a common design sequence. The first task of a system designer is the development of a workable basic concept. The second task is reducing the vulnerability of the system to failures of component parts, including human participants. The final task is balancing the benefits and risks of the new system, starting with the internalized economic costs. The external effects have rarely been analyzed, and it is only in recent decades that we have become deeply concerned with this difficult but important part of the design process.

Risks created by technical systems arise either from routine external effects considered acceptable at the time

Reprinted by permission from <u>Science</u>, vol. 208 (6 June 1980), pp. 1114-1119. Copyright 1980 by the American Association for the Advancement of Science.

of design, or from abnormal conditions that are not part of
the basic design concept and its normal operation. Most
abnormal events usually impair or stop the operation of the
technical system, and may threaten the operators. The usual
external effect is the loss of operational benefits to the
users of the output. The major internal consequences of
failures are borne by the operating institution. The timely
diffusion within the institution of information about such
failures usually stimulates rapid modifications to reduce
the ratio of failure costs to benefits. Less frequently, a
failure results in effects outside the institutional boundary, creating a public risk--and a potential cost to the
public. These external costs are usually difficult to evaluate, and here the informational mechanism for system modification is usually cumbersome and slow. In recent years
such modifications have been made because of an increasing
public concern over the inherent risks and costs arising
from previously acceptable external effects, both occasional
and routine. For these reasons, the importance of risk as
a design criterion is increasing.

The basic truisms about risk are readily recognized.
First, everyone knows that risk taking is an accepted part
of life. Living can be fun, but it is also dangerous (just
how dangerous can be difficult to measure). Second, everyone reacts differently to risks taken voluntarily and to
risks that are imposed by some outside group. Third, decisions imposing risks on us are being made all the time.
This results in the fourth truism: a conflict is inherent
when a group imposes a risk on others. Historically, such
conflicts have been resolved by compromise, but rarely to
everyone's satisfaction.

It is, therefore, characteristic of the functioning of
an organized society that conflicts arise from the balancing
of public benefits and involuntary risks to the individual.
Because such conflict is unavoidable, our problem is how to
manage and minimize it.

How should group decision processes operate to minimize
social costs and maximize social benefits? Group processes
range from anarchy to dictatorship. In most of the industrial world, we enjoy a medium between these, but the processes for decision-making have themselves become contentious
issues.

Social costs include intangibles, and the question
immediately evident is what costs are included and how are
they weighted. It is obvious that if we have a decision

process, and if we know how to determine costs to the individual, we still have a problem with the full disclosure of
all the social costs. What is full disclosure? Do we
include the options for societal risk management as part of
full disclosure: that is, the cost of the alternatives for
managing the risk? Does it include all present events,
future events, the people who get the benefits, and the
people who bear the costs? We have geographic distributions,
time distributions, demographic distributions--all of these
are included by the term full disclosure. Where do we draw
the boundaries?

Decisions are not made by institutions; the decision
process involves people. The government typically works
through agencies and committees, so that, in fact, it is a
few people in the agencies and a few people on the committees who really decide what happens. How do we allocate the
responsibility and the costs of bad decisions? How do we
functionally connect authority, responsibility for outcomes,
and costs?

After we establish the social costs, how do we set our
priorities? How do we determine the relative merits of
various outcomes? That is a subject for a separate study,
because, of course, value systems depend on culture, background, economic status, and all kinds of psychological
factors.

Part of the problem in risk assessment comes from confusions that arise during discussions of the subject--
confusions about reality, analysis, and individual perceptions. Reality is what has happened or what will happen.
Analysis is a process based on collected data, anecdotal
cases, and statistics, any of which may or may not be correct; and based on these, we invent simplified models to
predict an outcome. The result, of course, is a large uncertainty in the predictions.

What is the untuitive perception of the individuals
involved? Involuntary risks are perceived differently by
individuals. Their perceptions may be far from reality, so,
in discussing public acceptance of risk, we have to distinguish between the uncertain reality of what may occur, the
uncertain analysis of predicting it, and the variable perception of its potential. Similar confusions exist, in-
cidentally, over social costs and social benefits, which are
also involved. As an illustration, who in the year 1900
could have predicted the social costs and benefits of the
automobile?

Finally, people's perceptions of probabilities are
frequently in gross error. The accident at Three Mile
Island proved very little about probabilities of such events.
The inadequacy of such single events for providing probability numbers can be explained analytically, but the political
response and the public perceptions are often based on single
events. So even if a professional group develops analytic
answers, it has difficulty persuading the public to accept
them.

Recognizing all these difficulties, it is nevertheless
important to explore the subject of risk management in order
to improve the quality of decision-making.

Analytic and Judgmental Approaches

A question implicit in the term acceptable risk is
"acceptable to whom?" Certainly congressional approval of
any method for making risk-benefit decisions establishes its
legitimacy, but a public consensus is needed to sustain its
use. Defining this consensus is difficult because there are
technologies that are favored by a majority, or at least by
a plurality, but are opposed by extremely motivated individuals and groups (for example, those who fight water fluori-
dation and nuclear power). Because of our experience with
other political issues in which similar divisions of public
opinion occur (abortion, gun control), we know that we
should not be optimistic over the prospects that a regulatory approach can neutralize these controversies. Problems
such as these raise issues, such as the definition of majority versus minority rights and the scope and limit of due
process, that are well beyond those normally associated with
risk management.

Congress has not defined "acceptable" risk levels,
except for a few cases in which a zero risk approach was
mandated. Far more frequently,[1] Congress delegates responsibility for judging risk acceptability to regulatory agencies with the criteria that protection be provided against
"unreasonable" risks. The methods by which these agencies
interpret "reasonableness" range from a formal analysis of
risk, benefits, and alternatives to purely subjective evaluations.

Analytic Approaches

The attraction of analytic methods (cost-benefit analysis, decision analysis) is their capacity to make explicit
the assumptions, value judgments, and criteria used for

making a decision. The analytic approaches are considered
logically sound and sufficiently flexible to accept any
value system. Given a specific set of values and criteria,
a cost-benefit analysis could ideally indicate the decisions
that would best balance technological risk and benefit
(assuming that both tangible and intangible costs and benefits are included). But in reality it is difficult to
measure group values, and at best the analytic methods can
only be used to reach a rough approximation of the social
cost and benefits that characterize a decision.

The debate over the relative merits of these approaches
generally focuses on the effects of incomplete information
(omitted and uncertain risks, benefits, and values), neglect
of distributional effects, and other errors of simplifica-
tion. It is not our intent to review the merits of these
methods as commonly practiced; that has been done elsewhere.[2-6]

Physical Versus Financial Risk

Because of our use of the term risk as the probability
of either financial or physical damage, we may tend to allow
uncritically the use of premises about the acceptance of the
risk to "life and limb"[7] to be based on an analogy to financial risk taking.

From the societal viewpoint, the presumption that risk
equals cost may be valid in most cases. For example, the
cost of the risk of death is sometimes calculated as being
equal to the discounted net earnings of those killed. This
method, now out of favor, operates as if the loss of lives
were equivalent to the breakdown of productive machines.

Similarly, the value assigned to resilience[8] leads to a
desire to avoid catastrophic accidents that parallels the
strategy in which investments are diversified in order to
limit losses under adverse conditions. Perhaps recognizing
the differences between these two types of risk, Zeckhauser[9]
argued that, on a per fatality basis, the social cost of
multiple-fatality accidents is lower than that of a single
fatality because fewer survivors are affected. For example,
the social cost of the loss of a city or a family is less
than that of an equivalent number of independent, dispersed
fatalities. Although the basis for this argument is apparent, it is also incomplete. For example, it ignores the
value placed on the continuation of a family line; the
importance of this value is evident in the draft deferment
that was given to sole surviving sons. Similarly, Wilson[10]
noted:

Value of Risk Avoidance

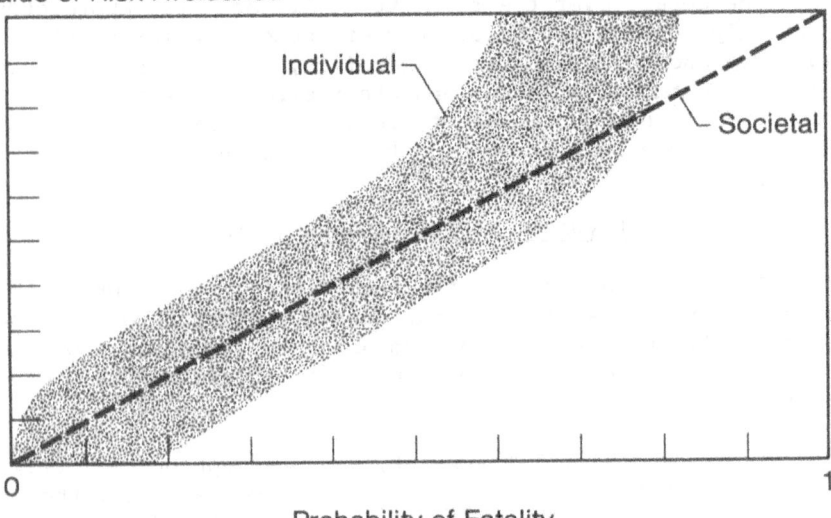

Probability of Fatality

Figure 14.1. Value systems for risk. Societal and individual risk-taking are not directly analogous, because individuals resist high risks, while societies frequently undertake activities with expected fatalities.

Small accidents throughout the world kill about 2 million people each year, or 4 billion people in 2000 years. This is "acceptable" in the sense that society will continue to exist, since births continually replace the deaths. But if a single accident were to kill 4 billion people, that is, the population of the whole world, society could not recover. This would be unacceptable even if it only happened once in 2000 years.

Another example in which the ability to generalize from financial cost-benefit analysis has been questioned is when physical risks are distributed across time. Arrow[11] argued that these risks should be treated as other costs, and dis-counted accordingly; other analysts of the issue have ques-tioned the validity of this approach, and looked for alter-native methods for guidance on how to judge equity in inter-generational risk trade-offs.[12]

Application of Expected Value to Individual Risk Assessment

Although the above analogy may be valid for the collec-tive view of the cost of risk, it may not apply to the in-tuitive evaluation of risk. As Fig. 14.1 illustrates, the individual and societal evaluations of risk are quite dif-ferent.[8] In the societal view, the presumption of a linear relation between risk and the cost of that risk may be quite valid. But as Howard[13] pointed out, the individual evaluation of that cost is necessarily nonlinear and becomes infinite as the probability approaches unity.

The use of an expected value or expected utility model is based on the premise that expected cost is simply the product of the probability of an outcome and the evaluation of that outcome. But in the individual's view of the risk of death, this is not valid, for this product is very large or infinite.

It is often presumed that the individual evaluation of the cost of risk is linear over a probability range of interest,[9,13] but there is little firm evidence to support any hypothesis about the shape of this curve, as far as we know. Under the common convention of risk analysis, the slope of the curves in Fig. 14.1 is referred to as the value of life. The politician's old saw that life is of infinite value can be reconciled if this refers to the individual evaluation of one's own life. This viewpoint is not incon-sistent with the assignment of finite costs to risks; it is the application of an expected value model that is inappropriate for this evaluation.

A second drawback with the application of expected value to individual risk evaluation stems from the tendency by analysts to seek to accommodate differences of opinion entirely within the assignment of utility.[14] This is because in most decision or cost-benefit analyses, the probability estimates are considered roughly valid because they are based on available data, engineering models (such as fault trees), and expert opinion. But in a study of public attitudes about nuclear power, the bulk of the disagreement was found to be due to different beliefs about accident probability.[15] Although it may be perfectly valid to base public policy on expert estimates and data, the attempt to reconcile differences in the assignment of costs and values is misdirected if, in fact, the controversy over technological risk is due to divergent beliefs about probability.

Intuitive Versus Analytical
Risk Assessment

We now consider the implications of the premise that risk acceptance is ultimately inseparable from the psychology of risk perception and evaluation. A corollary to this premise is the assumption that when the results of intuitive risk assessments differ significantly from those of the analytical methods, conflict follows.

It seems clear that intuitive and quantitative risk-benefit assessments can produce quite different results, even given the capacity of the analytical approaches to accommodate complex values relating to different risk attributes. The differences of opinion over probability assignments are not limited to those risks for which data are not available; many people intuitively fear travel by airplane more than by automobile, yet aviation is safer. Explanations of this effect focus on the degree of individual control over risk,[16] the conditional probability of survival given an accident, and the catastrophic nature of airplane accidents.[17]

The difficulty that arises from these differences in assessment stems from the dual meaning of acceptable risk. The analytical methods help regulators set standards that implicitly define acceptable risk. But the intuitive individual assessments of acceptability can overrule these decisions through the political process. The repeal of the seat belt interlock regulation and recent congressional action to prevent a ban on saccharin are cases in which public opinion resulted in a policy change.

Intuitive Risk-Benefit Analysis

Given the role of individual judgments of (physical) risk and benefit in determining the political acceptability of specific technologies, it seems particularly valuable to try to understand intuitive risk-benefit analysis. Efforts to develop this understanding were made by Starr,[8,18,19] whose approach was based on a study of historically accepted risk (revealed preferences) and by Fischhoff et al.[17] and Slovic et al.,[20] whose approach was usually based on risk-taking behavior as determined by questionnaire (expressed preferences). An additional source of information is the study by Lawless[21] of many controversies over technology. If we assume that many of these controversies arose because of intuitive estimates of unreasonably high risk (not true in all the cases described; some cases, such as the thalidomide tragedy, were due to late identification of a risk), then the common characteristics of risk and benefit in these controversies may indicate important factors in the intuitive risk process. Lawless did this, and his findings confirm those of other studies in identifying catastrophic potential and lack of individual control over risk as "factors that influence the impact of the threat."

Understanding the Intuitive Process

There is an attraction to try to develop an understanding of intuitive risk-benefit decisions by constructing parallels to the analytical methods. This approach leads to a model of intuitive decision-making in which subjective judgments of the probability and consequence of undesirable outcomes are somehow combined to produce a perceived risk; parallel judgments provide a perceived benefit; the two are then compared to provide intuitive judgment of acceptability.

This model is quite broad; it does not specify the intuitive procedures for arriving at either perceived risk or benefit, or for their comparison. Even so, the available evidence suggests that this model may be incorrect. First, studies of intuitive decision-making in general (not limited or applicable to physical risks alone), have identified numerous decision-making rules that do not follow the model described above.[22] Second, there is evidence to indicate that benefits are not intuitively evaluated independently from risks.

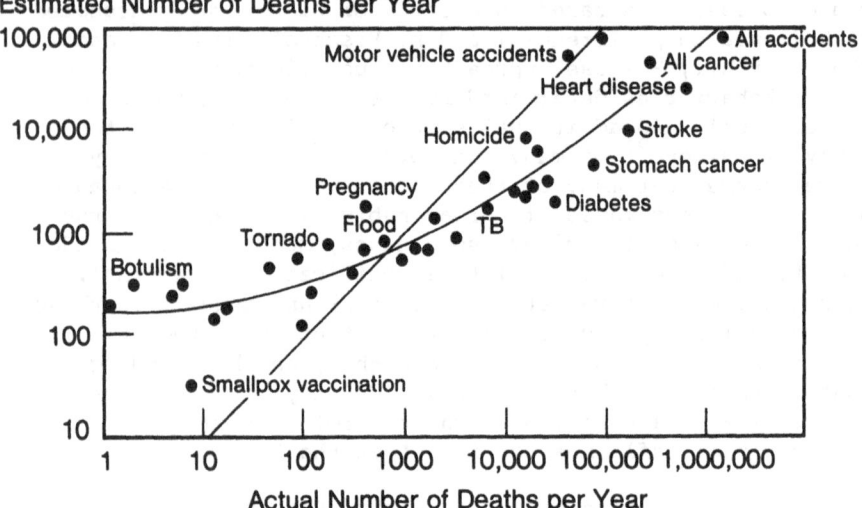

Figure 14.2. Comparison of perceived risk with actual risk. Reprinted by permission from S. Lichtenstein et al., "Judged Frequency of Lethal Events," Journal of Experimental Psychology, vol. 4, pp. 551–578. Copyright 1978 by the American Psychological Association.

In the survey of subjective risk and benefit by
Fischhoff et al.,[17] perceived risk and perceived benefit
were negatively correlated, due principally to the subjective evaluation of a number of things as high in risk and
low in benefit (handguns, cigarettes, motorcycles, alcoholic
beverages, nuclear power). When subjects were asked to
judge "the socially acceptable level of risk," those who
first took the benefits into consideration consistently
reported higher levels of acceptability than did subjects
who first evaluated risk, which reinforces the view that
risks and benefits are not evaluated independently.

Despite the limitations of the perceived risk-perceived
benefit view of deciding risk acceptability, we know of no
better way to attempt to understand the intuitive processes
for risk decisions. Support for this approach stems from
the fact that the acceptability of a risk has been found to
increase with increasing benefit both by Starr[18] and
Fischhoff et al.[17]

Benefits

Little work has been done to characterize the perceived
benefits of technological activities. Starr[18] found a
correlation between risk and "benefit awareness," which he
described as a crude measure of public awareness of social
benefits. This measure was based on the relative level of
advertising, the percentage of the population involved in
the activity, and a subjective judgment of the usefulness of
the activity. The survey by Fischhoff et al.[17] included a
subjective ranking of benefits, but no attempts were made to
relate perceived benefit with any characteristics of that
benefit.

Probability Perception

By far the most studied and best understood component
of intuitive risk benefit analysis is risk perception and
evaluation. There is excellent literature on the subjective
estimation of probability.[23]

One aspect of the interpretation of probability that
has been noted repeatedly is the intuitive handling of very
low probabilities. As Mishan[7] noted: "One chance in 50,000
of winning a lottery, or having one's house burned down,
seems a better chance, or greater risk, than it actually
is." The same observation was made by Selvidge.[4] Lichtenstein et al.[25] found similar results (Fig. 14.2) when they
asked people to estimate the number of fatalities from
specific causes annually in the United States: "The full

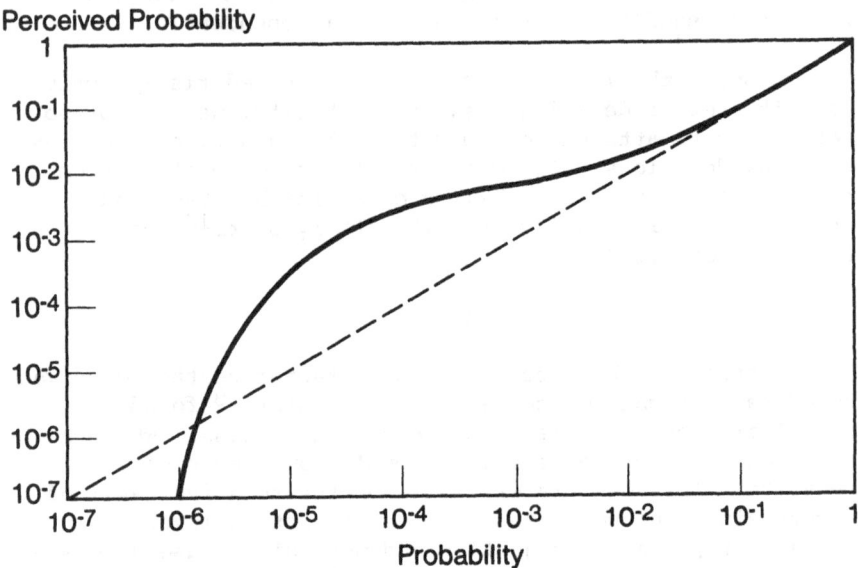

Figure 14.3. Perception of probability. We postulate
that the compression of perceived probability (as illustrated
in Figure 14.2) does not extend to zero risk. At some point
risk interpretation goes from "low" to "negligible."

range of perceived risk is only about 10,000 while the corresponding actual range is closer to 1,000,000." Similar results were found in another survey in which risk was ranked subjectively.[17]

The influence of this perception is important when we recall that the expected value or expected utility model calculates that a change in event probability by a factor of 1000 produces a change in expected value or utility by a like amount. If the probability is perceived as having changed by a much smaller amount, then it would not be surprising to find that an intuitive evaluation of risk is less sensitive to probability changes. This can be extremely important for low-probability, high-consequence risks, because probabilities lying below an intuitively understandable range may be overestimated.

We postulate that this is only true to a point (see Fig. 14.3). Although we selected these scales judgmentally, their chief purpose is to illustrate that, at some low level of probability, the intuitive interpretation goes from "low" or "unlikely" to "negligible" or "impossible." This hypothesis can be used to explain behavior regarding seat belt use, and perhaps smoking. In a study of seat belt use, Slovic et al.[26] noted that if the decision to wear seat belts is approached on a per trip basis, "we might expect that many motorists would find it irrational to bear the costs (however slight) of buckling up in return for partial protection against an overwhelmingly unlikely accident." They observed that "change of perspective, towards consideration of risks faced during a lifetime of driving, may increase the perceived probabilities of injury and death and, therefore, induce more people to wear seat belts....Such differing perspectives may trigger much of the conflict and mutual frustration between public officials and motorists, each believing (with some justice) that their analysis of the situation is correct." Similarly, Jacobson[27] referred to carcinogenic "chemicals which pose minuscule hazards to individuals, but significant hazards to the population as a whole." This last point supports our premise that much conflict over technological risk is due to differences between intuitive and analytical risk-benefit analyses. If the hypothesis that perceived probability is effectively zero for some risks is valid, then the perceived risk of a short automobile trip without seat belts or of one cigarette may be zero.

This nonlinearity in probability perception indicates that even something apparently as basic as the unit of exposure used to evaluate risk can be influential. In his analysis, Starr[18] commented, "The hour-of-exposure unit was chosen because it was deemed more closely related to an individual's intuitive process in choosing an activity than a year of exposure would be."

Accepting, at least tentatively, the relation between perceived and actual probability (Fig. 14.3), we can see a basis for the controversy over catastrophic risks. As mentioned above, high-consequence, low-probability risks are of particular concern if their probabilities are overestimated subjectively. But when part of the public believes the probability is low and another part believes it to be negligible, these beliefs lead to radically different evaluations. This may be the case with nuclear power and other risks of this type, and may be a key reason for the controversies over these risks.

Risks Distributed Over Time

Given the apparent nonlinearities in risk evaluation depending on the unit of measurement, it seems reasonable to look for other perceptual factors related to the units in which risks are expressed. A number of distinctions can be considered: risks can be immediate or delayed, cumulative or ephemeral, and can affect future generations or our own or both. There is little evidence to indicate how these factors are handled. Fuchs[28] cited evidence that individual discount rates for financial and physical risk are positively correlated. But the fact remains that benefits and risk may be discounted at different rates. For decisions with very long-term implications, the use of a variable discount rate, declining with time, may more accurately reflect the value given to future risks and benefits than a constant discount rate.[29] This is an area that seems particularly worthy of attention, for many risk controversies are about risks that are persistent or cumulative, such as carcinogens.

Predicting Risk Controversies

Because of the work to define the factors influencing perceived risk, it is now possible to anticipate the kinds of risk likely to generate controversy. Catastrophic potential and lack of individual control, particularly once an accident occurs or a risk is identified, are apparently the most important risk characteristics. When the uncertainty associated with risks is great, data concerning the uncertainty not forthcoming, and expert opinion apparently divided, apprehension by the public is understandable. Haefele[30] termed these risks hypothetical and described nuclear power as the "path-finder" for these risks. Certainly there are many risks with the characteristics described above (for example, toxic chemicals and recombinant DNA research). Whether decisions can be made about these risks without the high degree of controversy and the resulting high social cost associated with the nuclear debate remains to be seen.

Quantitative Criteria for
Risk Acceptance

In May 1979, the Advisory Committee on Reactor Safeguards (ACRS) recommended "that consideration be given by the Nuclear Regulatory Commission (NRC) to the establishment of quantitative safety goals for overall safety of nuclear power reactors."[31] The ACRS further recommended that "Congress be asked to express its views on the suitability of such goals and criteria in relation to other relevant aspects of our technological society...." A similar suggestion, accompanied by proposed criteria, was made by Farmer[32] in 1967; the criteria were expressed by a curve relating acceptable accident frequency with accident magnitude. Subsequent proposed criteria for acceptable risks, not necessarily limited to nuclear power, have been made by Starr,[18] Bowen,[33] Rowe,[34] Okrent and Whipple,[35] Wilson,[36] and Comar.[37] Currently efforts are under way within the NRC, the ACRS, and elsewhere to develop quantitative criteria for risk acceptance and to consider the many issues raised by this approach.

Incentives to Develop Quantitative
Criteria for Acceptable Risks

The dissatisfaction with current regulatory systems for risk management provides impetus to develop new methods. Theoretically, quantitative criteria for acceptability would resolve many specific criticisms. One criticism stems from the fact that in several cases, a zero risk goal has been established. This denies the concept of a trade-off between risk and benefit, and ignores the difficulty or impossibility of reaching zero risk. Further, improvements in technology have permitted identification and estimation of risk at levels far below those that were possible when specific zero-risk laws were passed; risks we might consider negligible are not treated in the regulatory process differently from much higher risks. As Hutt[38] argued,

Until quite recently, a no-risk food safety policy was widely thought to be an achievable goal....It is now clear that it is literally impossible to

eliminate all carcinogens from our food. Moreover, many of the substances which pose a
potential risk are part of long-accepted components of food, and any attempt to prohibit
their use would raise the most serious questions
both of practicality in implementation and of
individual free choice in the marketplace.

A suggested way of handling this problem would be to set a
level below which risks would be ignored, provided some
benefit were associated with the risk. This low level would
serve as a quantitative standard for acceptability of the
risk.

A second criticism of regulatory approaches is that
decisions are often made arbitrarily. Such a charge is not
surprising considering that several regulatory agencies have
a mandate to protect the public from "unreasonable" risk,
without congressional guidance on how to judge reasonableness. The objections are enhanced when regulators are
believed to be overly accommodating or hostile to the regulated industry. Certainly, one way to reduce the influences
of bias and arbitrariness is to institute a numerical definition of "reasonable." Perhaps the time required for risk
decisions would also be reduced by the availability of clear,
relatively simple criteria.

Often, regulatory authorities specify the technology
for meeting risk targets, rather than the targets themselves. The drawback of this approach is that there are
no incentives to develop more efficient methods of controlling risk. The establishment of risk targets alone could
stimulate the development of a variety of creative methods
for risk control.

Finally, another criticism of current risk management
is that the effort required to control risk (as measured by
the cost per life saved) varies considerably from one risk
to another; this wastes both lives and money.[9] Assuming
that the total funds allocated for risk reduction could be
transferred freely between different risk reduction opportunities (which is certainly not always possible), the
maximum number of lives that could be saved nationally is
found when the marginal cost of saving a life is uniform
among the opportunities. Thus the comparative marginal
cost-effectiveness of each opportunity for saving lives
would become the guiding principle in the allocation of
resources, and the value of life would be implicit in the
total national allocation of funds. There would, of course,
need to be a national allocation of resources to such "life-

saving" endeavors, but as with military budgets, a common-sense consensus judgment is likely to be as reliable as any analytic formula.

Applications for Risk Criteria

One of the pitfalls in trying to develop regulatory approaches for managing risk is the desire to use the same method to tackle a number of different risks. There are different types of risk decisions, and no single regulatory method seems applicable to all of them.

The use of cost-effectiveness criteria serves as an example. This issue arises when, a priori, the technology is found acceptable but the specific operating point is left to be decided. An example of this type of decision is the determination of allowable levels of a pollutant in automobile exhaust. In this case the issue involved is not the relative risk and benefit of transportation, nor the selection of a transportation technology (automobile versus mass transit). For this simplified type of decision the only issue is the marginal trade-off between the social cost of the risk and the cost of controlling it. For these cases, two kinds of quantitative criteria can be considered; the first is the standard for judging cost-effectiveness described above. There is nothing new in this approach; it is simply cost-benefit analysis in which the metric for judging the social cost of risk has been specified. The second quantitative criterion is more pragmatic: it is a lower risk limit below which no regulatory action would be taken. This could be useful in allocating regulator's time and would help prevent the highly visible cases in which the nuisance aspects of regulation are intuitively greater than the benefits of that regulation.

The next level of difficulty in risk decisions is the choice of the best method for obtaining a specific benefit. In these cases the benefits need not be analyzed, it is presumed that the benefits are sufficiently great to justify any of several alternatives. For example, in the often heated debates over the selection of energy production technologies, it is generally assumed that, under any proposed policy, energy services will be provided (such services include conservation). For this decision, the dominant issues are the costs and risks associated with each alternative. The difficulty in making these choices is often due to the qualitatively dissimilar character of the risks (for example, air pollution risks from coal mining and burning versus nuclear reactor accident risks). It is difficult to see a role for quantitative criteria in making

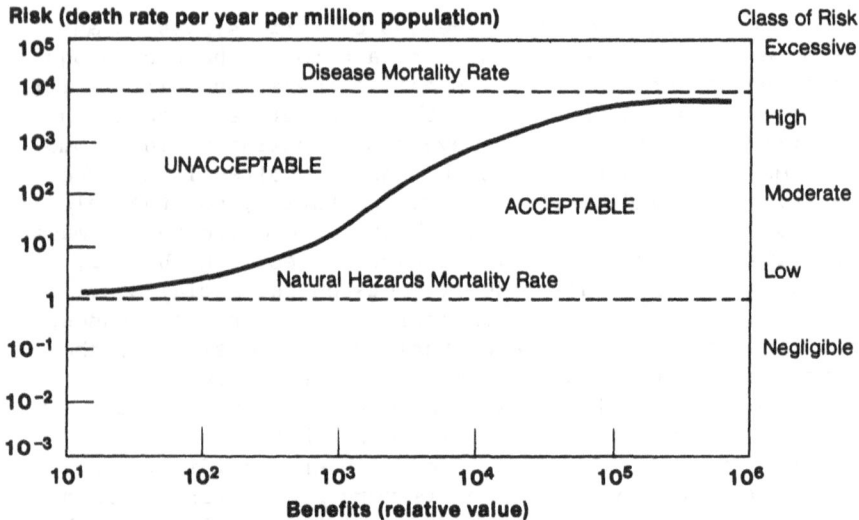

Figure 14.4. Risk/benefit pattern for involuntary exposure. Proposed risk acceptability criteria frequently follow this form. There is a lower limit of concern, an upper limit for acceptability, and criteria for risk-benefit trade-off between these limits.

comparisons of the type needed. One could establish a
maximum permissible risk level that would serve to screen
out excessively risky alternatives, but the selection of a
technology generally depends, either explicitly or implicitly, on some aggregation of the components of the social
cost of each alternative. Presumably, after one alternative
is selected, the decision is reduced to the determination of
the preferred operating point discussed above.

A complete risk-benefit decision requires that the relative social cost of the risk be compared with the associated benefit. A pragmatic application of quantitative criteria for these cases was suggested by Starr[19] and by Starr
et al.,[8] and is illustrated in Fig. 14.4. This risk-benefit
curve reveals the commonly proposed characteristics for
risk criteria: a lower limit for concern about risk (in this
case, the natural-hazards mortality rate), an upper limit
for acceptability (set by the average disease rate), and provision for risk-benefit trade-off between these limits.

We should not overestimate the capacity of simple criteria, such as those illustrated in Fig. 14.4, to reduce
risk conflict costs. Many, if not most, risk estimates
include significant judgmental inputs. There are often
substantial disagreements over risk estimates, the method
used to arrive at risk estimates, and the competency, inte-
grity, and motivation of the experts providing subjective
risk estimates. What is needed for the application of quantitative criteria for risk acceptance is a standard of proof
for determining whether the criteria have been met. Although many different approaches to this issue have been
recommended (including peer review, scientific courts, and
quantitative methods for resolving differences between
experts), the ultimate responsibility for judging the competency of risk analysis still resides with the regulatory
agency responsible for managing the specific risk.

Conclusion

Analytical approaches to decide risk-benefit issues
ideally come closer to maximizing net social benefits than
any other approach. The usefulness of these methods in
making assumptions and values explicit justifies their application. But a necessary condition for applying their results
to specific decisions is a social consensus on the relative
benefits and costs of the proposed actions. For specific
types of risk, in which intuitive evaluations of risk and
benefit contradict analytical evaluations, the necessary
consensus may not develop, but rather a conflict requiring
political resolution is likely to result.

When the conflict arises from a disagreement over the level of risk rather than the value assigned to that risk, efforts to reduce the cost of conflict by incorporating values into an expected utility approach will be unsuccessful. Quantitative risk criteria appear quite attractive in this respect, because the key to the acceptability of a technology under the proposed method is the level of risk. Assuming that the estimated risk became the central point in the debate, the public might have more confidence in the regulatory systems if their concern were directly addressed.

We see significant value in trying to understand the intuitive risk-benefit process. The evaluation of its outcome could reduce anxiety and cost if used as a tool in the design of technical systems. This is already the case, as when we use more stringent criteria for nuclear power and commercial aviation than for a more commonplace risk.[39,40] The balance between individual and group risk-benefit decision methods is fundamental to the development of national policies on risk acceptance. It is customarily achieved through the political process and is not amenable to quantitative analysis.

References and Notes

1. W.W. Lowrance, *Of Acceptable Risk* (Kaufman, Los Altos, Calif., 1976).

2. Mitre Corporation, METREK Division, *Symposium/Workshop on Nuclear and Nonnuclear Energy Systems: Risk Assessment and Governmental Decision Making, Proceedings* (MTR 79W00335; Mitre Corporation, McLean, Virginia, September, 1979).

3. L. Lave, "Risk Assessment and Cost-Benefit Analysis," in Ref. 2, pp. 176-186.

4. S. Jellinek, "Risk Assessment at the Environmental Protection Agency," in Ref. 2, pp. 59-64.

5. M. Green, "The faked case against regulation," *Washington Post,* 21 January 1979, p. C-5.

6. P.H. Schuck, "On the chicken little school of regulation," *Washington Post,* 28 January 1979, p. C-8.

7. E.J. Mishan, *Cost-Benefit Analysis* (Praeger, New York, 1976).

8. C. Starr, R. Rudman, and C. Whipple, "Philosophical Basis for Risk Analysis," *Annual Review of Energy,* 1 (1976), 629-662.

9. R. Zeckhauser, "Procedures for Valuing Lives," *Public Policy,* 23, No. 4 (1975), 419-464.

10. R. Wilson and W.J. Jones, *Energy, Ecology, and the Environment* (Academic Press, New York, 1974).

11. K. Arrow, "The Rate of Discount for Long-Term Public Investment," in *Energy and the Environment: A Risk-Benefit Approach,* ed. H. Ashley *et al.* (Pergamon, Elmsford, New York, 1976), pp. 113-140.

12. National Planning Association, *Social Decision-Making for High Consequence, Low Probability Occurrences* (PB-292735; National Technical Information Service, Springfield, Va., 1978).

13. R. Howard "On Making Life and Death Decisions," in *Societal Risk Assessment: How Safe is Safe Enough?,* ed. R.C. Schwing and W.A. Albers (Plenum, New York, 1980), pp. 89-106.

14. W. Haefele, "Benefit-Risk Trade-Offs in Nuclear Power Generation," in Ref. 11, pp. 141-184.

15. H. Otway, J. "Risk Assessment and the Social Response to Nuclear Power," *British Nuclear Energy Society: Journal,* 16 (1977), 327.

16. S. Montague and E. Beardsworth, "Benefit/Risk--A Critique, and Cultural Analysis for Non-Quantitative Risk-Assessments," unpublished manuscript (n.d.).

17. B. Fischhoff, P. Slovic, S. Lichtenstein, S. Read, B. Combs, "How Safe is Safe Enough? A Psychometric Study of Attitudes towards Technological Risks and Benefits," *Policy Sciences,* 8 (1978), 127-152.

18. C. Starr, "Social Benefit Versus Technological Risk," *Science,* 165, (1969), 1232-1238.

19. C. Starr, "Benefit-Cost Studies in Sociotechnical Systems," in *Perspectives on Benefit-Risk Decision Making* (National Academy of Engineering, Washington, 1972), pp. 17-42.

20. P. Slovic, B. Fischhoff, S. Lichtenstein, "Facts and Fears: Understanding Perceived Risk" in *Societal Risk Assessment: How Safe is Safe Enough?*, ed. R.C. Schwing and W.A. Albers (Plenum, New York, 1980), pp. 181-214.

21. E.W. Lawless, *Technology and Social Shock* (Rutgers University Press, New Brunswick, N.J., 1977).

22. I. Janis and L. Mann, *Decision Making* (Free Press, New York, 1977).

23. See, for example, A. Tversky and D. Kahneman, "Judgment Under Uncertainty: Heuristics and Biases," *Science*, 185, (1974), 1124-1131.

24. J. Selvidge, "Assigning Probabilities to Rare Events" (thesis, Harvard University, 1972).

25. S. Lichtenstein, P. Slovic, C. Fischhoff, M. Layman, B. Combs, "Judged Frequency of Lethal Events," *Journal of Experimental Psychology: Human Learning and Memory*, 4 (1978), 551-578.

26. P. Slovic, B. Fischhoff, S. Lichtenstein, "Accident Probabilities and Seat Belt Usage: A Psychological Perspective," *Accident Analysis and Prevention*, 10 (1978), 281-285.

27. M.F. Jacobson, letter in *Science*, 207, (1980), 258-261.

28. V. Fuchs, "The Economics of Health in a Post-Industrial Society," *Public Interest*, 56 (1979), 3.

29. J.M. English, "A Perceptual-Time Scale for Determination of a Discount Function," in *Trends in Financial Decision Making*, ed. C. van Dam (Nijhoff, The Hague, Holland, 1978), pp. 229-247.

30. W. Haefele, "Hypotheticality and the New Challenges: The Pathfinder Role of Nuclear Energy," *Minerva*, 10 (July, 1974), 314-315.

31. M.W. Carbon, letter to J.M. Hendrie, 16 May 1979.

32. F.R. Farmer, "Siting Criteria--A New Approach," *Containment and Siting of Nuclear Power Plants* (International Atomic Energy Agency, Vienna, 1967), pp. 303-318.

33. J. Bowen, "The Choice of Criteria for Individual Risk, for Statistical Risk, and for Public Risk," in *Risk-Benefit Methodology and Application: Some Papers presented at the Engineering Foundation Workshop, September 22-26, 1975, Asilomar, California*, D. Okrent, ed. (Rep. PB-261920; School of Engineering and Applied Science, University of California, Los Angeles, 1975), pp. 581-590.

34. W.D. Rowe, *An Anatomy of Risk* (Wiley, New York, 1977).

35. D. Okrent and C. Whipple, *An Approach to Societal Risk Acceptance Criteria and Risk Management* (Rep. UCLA/ENG-7746; School of Engineering and Applied Science, University of California, Los Angeles, 1977).

36. R. Wilson, testimony before Occupational Safety and Health Administration hearings (OSHA-docket H-090; Washington, February, 1978).

37. C.L. Comar, "A Pragmatic de minimis Approach," *Science, 203*, (1979), 319.

38. P.B. Hutt, "Unresolved Issues in the Conflict Between Individual Freedom and Government Control of Food Safety," *Food, Drug, Cosmetic, Law Journal, 33* (1978), 558.

39. C. Sinclair, P. Marstrand, P. Newick, *Innovation and Human Risk* (Centre for the Study of Industrial Innovation, London, 1972).

40. R. Schwing, *Expenditures to Reduce Mortality and Increase Longevity* (General Motors Research Laboratories Report GMR-2353; Warren, Michigan, 1977).

15. Dealing with Industrial Health Risks: A Stepwise Goal-Oriented Concept

From the earliest of times, industrial activities of all types have used, produced, and disposed of a wide variety of natural, and more recently, synthetic materials. Some of these materials have been found in recent times to induce serious, chronic health effects in workers, consumers, and members of the general public.[1] More and more, scientific research supports a correlation between chronic health effects and past exposures, particularly in the area of cancer,[2] and people may be exposed to chemicals in a variety of ways. Elsewhere in this volume, for example, Librizzi (Chapter 6) and Ember (Chapter 7) have discussed the highly publicized episode at Love Canal in which industrial wastes disposed of in accordance with the laws and practices of an earlier time have, through a series of later contretemps, emerged to cause exposure to unsuspecting people. The Thomas report[3] points out that the real extent of the chronic health effects, if any, has not yet been determined, but the incident serves to illustrate one of the many ways in which exposures can occur. And Coates (Chapter 2) has pointed to a thousand possible risks in the technological society.

The problem for industry and government today is to recognize, to assess, and to reduce the risks that such diverse phenomena pose, eliminating risks where possible--all in the face of limited resources of all kinds, limited information, and uncertainty in the scientific data and even in the total magnitude of the problem. One especially critical resource is the number of diverse, qualified, scientifically or medically trained people available.

Uncertainty is inherent in existing toxicological and epidemiological data.[2,4,5,6] More often than not one must

depend on toxicological data from animal experiments which
are not only error-prone by nature, but difficult to interpret in human terms. Results are usually obtained at dosages
yielding disease incidence levels far above those relevant to
human safety, because statistical significance with a reasonable number of test animals cannot be achieved otherwise. At
such high dose levels observed phenomena may not be biologi-
cally or mechanistically relevant, and their extrapolation to
low doses adds further uncertainty even if the problems of
translating animal test results to human terms have been overcome or ignored. The problems of uncertainty are well-known
to be extremely severe even in the relatively well-known case
of cancer (Gori, Chapter 11).

As a result of uncertainty, we do not have today a generally agreed upon understanding of what the real contribution of industrially related cancer is to the total incidence
of cancer. Some of the best studies done to date indicate
that from 1 to 5% of all cancer is occupationally related.[7-13]
Another[14] gives a somewhat higher figure, and one,[15] which
has been justly criticized,[16-19] proposes that at some point
in time the figure may range from 23 to 38%. It is doubtful
if any detectable portion of the cancer burden can be attributed to exposure to industrially derived agents in nonoccu-
pational settings,[19,20] though this field, too, is an area of
controversy.

Measuring the ultimate effects of controls some years
after they have been put into effect requires at least that
we be able to determine how much cancer is caused by industrially related exposures. With the evidence currently
available, this is something we cannot yet do; and if, as
seems likely, the contribution is near the 5% range or so,
the effect of control measures may be impossible to
see.[13,19-21] To this difficult situation must be added a
regrettable amount of politicization.[22,23]

Whatever policies and regulatory systems emerge for
controlling industrially related chronic health effects,
they must meet a twofold objective: (1) to cause deployment
of our resources and information so as to protect the most
people from potential hazards as soon and as effectively as
possible; and (2) to lay a foundation for continuing risk
reduction and control. Moreover, this must be done in the
face of the many large uncertainties. The following discussion focuses on cancer because it is in many ways a typical,
serious, chronic health hazard with relation to exposure to
industrial materials. It is also the chronic health problem
of greatest current interest.

The Problem of Cancer Prevention:
Present Situation

Cancer is the second leading cause of death in the United States. It has been widely publicized that something like one person in four will at some time in life contract cancer from one source or another. Even though the etiology is unsettled in many respects, we know that industrial activity is a contributor to this incidence. Whatever the actual contribution of industrially derived agents may be, industry must work vigorously to solve the problem. In addition many governmental agencies are now attempting to frame, under appropriate legal mandates, policies and regulations on the issue.[24-28] Thus, the Food and Drug Administration was the first agency to have made use of quantitative risk analysis in the area;[6] and the Occupational Safety and Health Administration (OSHA) has recently issued regulations.[29]

Where policies and regulations are sound, the talent and money to implement them are well-spent. But policies and regulations may be arbitrary because *a priori* policy decisions have been made which leave at risk people who, through another approach, might have been saved from contracting cancer. This may result from failure to call for the scientific best estimation of risk as a centralizable activity distinct from regulatory decisions. So far the only governmental agency that has proposed this distinction is the Office of Science and Technology Policy,[30] which itself has no specific regulatory function.

A generic problem facing both industry and government is how to set exposure levels which are considered by one criterion or another to be "safe." This raises the well-known question of "How safe is safe?" and the implied problem of trade-off between risks and benefits.[31] Because of the recognition of various types of benefits, federal agencies have generally recognized that outright banning of industrial carcinogens cannot solve the problem. Banning can, indeed, engender other serious risks. In place of banning, two other approaches have been put forward.

One such approach has been to set levels of risk so low that no "real" risk is run by exposed individuals. This amounts to establishing a risk floor at or below which the risk is considered "fully acceptable." Lifetime probabilities of one in 10,000,000 to one in 100,000 have been suggested for this floor.[32-35] Quantitative risk assessments at such levels unfortunately involve extrapolating exposure-response relationships by many factors of ten in exposure. The reliability of such extrapolation is very low and when

deduced exposure limits are translated into resources required to achieve them, the meaning becomes clear. Thus, an
error of a factor of 10 or 100 in predicted exposure can
spell the difference between a resource demand that is reasonable and one that simply cannot be met. One possible
consequence of insisting on "near-zero-risk" policies is
severe social and economic disruption without concomitant
benefits to health. This suggests that one ought to explore
other approaches that do not stretch quantitative risk
assessment quite so far.

A second way of defining an acceptable level of risk is
based on the idea that once a substance is classified a carcinogen, exposure is to be lowered to the lowest "feasible"
level.[24,29] This does not, on the face of it, involve making
a quantitative risk assessment. What is feasible, however,
is usually left undefined and may generate prolonged debate
and conflict, leading to the same counterproductive results
as the approach of near-zero risk ceilings.

One of the most undesirable consequences of these policies is that through conflict they cause intense focusing of
major effort on a single or a very few substances. This can
delay recognition, assessment, and some degree of control
over the full range of potential carcinogens. The reports
we read in our newspapers,[35] the results of the National
Cancer Institute's bioassay program,[36] and many other findings[37] and reports indicate that the number of potential
carcinogens may be found to be large. Should this be so,
there is a need to attack the problem from new perspectives
(see Chapters 12 and 15) and to examine policies that have
the explicit objective of maximizing effective cancer prevention and control.[38]

The remainder of this article describes one such approach. It involves an alternative way to use quantitative
risk assessment, a definition of feasibility and when to use
it, and the prospect of *more* than achieving, over time, a
quantitative national goal for industrially related cancer
prevention.

The Problem of Cancer:
An Alternative Solution

If risk-floor policies defining levels *below* which risk
is certainly acceptable do not solve the problem, does it
make sense to seek risk-ceiling policies specifying levels
above which, even for high benefits, risk is certainly
unacceptable? Because there is much room to argue when

defining acceptability, it is possible that the latter offers significant hope. Ceilings, as defined, will lead to a fundamentally different attack on the problem, especially when considerations of aggregate risk are included: the goal of policy should not merely be reduction of cancer incidence for each individual carcinogen, but rather to affect the total risk of cancer associated with each kind of exposure situation.

A homely example, woodcarving, may serve as partial illustration of the suggestion to be made. In carving a wooden dog using only a woodcarver's knife, one could take the tack of first carving an ear as exactly as possible in what is judged to be the right location on the piece of wood; next, of carving a second ear, or a tail, and so forth until the job is complete. If all goes well, after expending much time and talent, the result may resemble a dog, though individual parts can be badly out of place. The more usual way of woodcarving involves roughing out the general form of the dog, and then improving on this form, once the overall relationships are in place. Following this route, more wood is carved sooner and even when no single feature of the dog is yet complete, the eventual result may be recognized early in the process. Further, in this process the carver may learn from what he has done, and in this way assure that the final result will indeed be a carving of a dog, even if individual parts are not carved to perfection.

The near-zero risk approaches to the problem of cancer prevention resemble the first woodcarving technique, whereas what is needed is something like the second. The principal feature of the second approach is the setting of an interim, broad-scope goal that in a relatively short time provides significant protection to a large number of people, and then is in time further refined as knowledge, methodology, and technology improve. The use of the concept of setting a goal for cancer risk reduction is described below.

It is worth noting just what is meant by the idea of "risk ceiling." To illustrate, Figure 15.1 shows a series of words describing different intensities of risk and covering a range from "unacceptable" at the high end to "zero risk" at the low end. Lying between are "substantial," "significant," and "not detectable" risks. With reference to Figure 15.1, the level C would constitute a risk ceiling, and the level B a risk floor below which, in a strict sense, quantitative assessments are of very dubious validity. The "near-zero" risk floors discussed earlier are equivalent to level A, and are separated from the practical floor, B, by

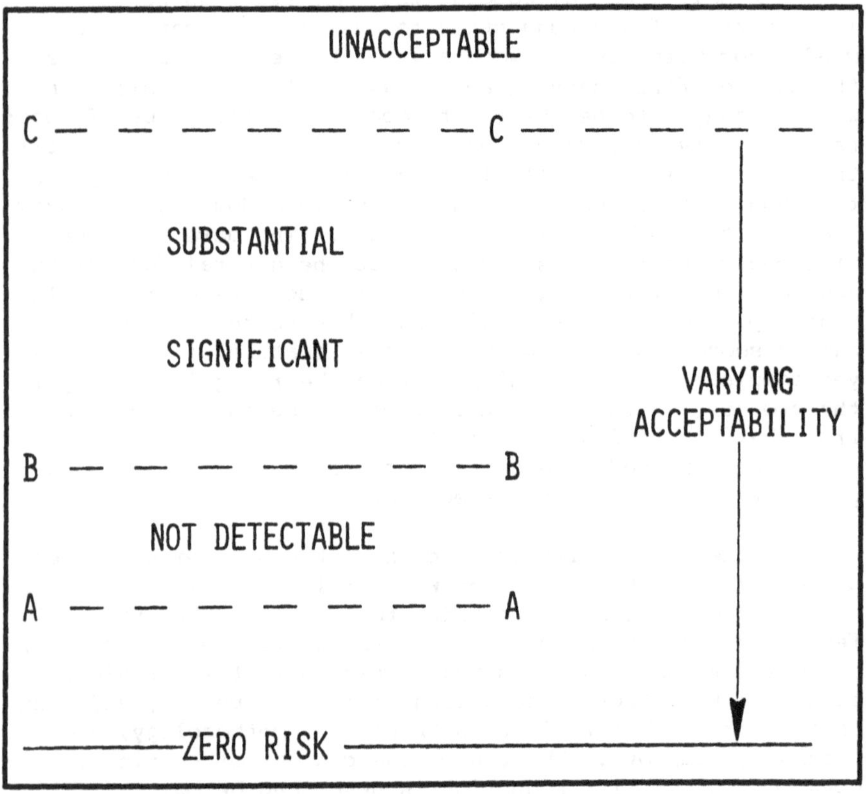

Figure 15.1. Qualitative societal health-risk categories.

a region in which effects are not detectable and in which, therefore, further reductions in exposure are not likely to produce significant further improvements in human health.

Risk ceilings should apply to exposure situations, not merely to individual carcinogens. Considering an individual person, the kinds of exposure situations that an individual can experience include: the air breathed, the water supply, the food supply, the medical/drug supply, numerous consumer products, and the work environment. People thus encounter several different kinds of exposure situations, and the risk of contracting cancer must be related to the summed effect of these exposures. A policy based on risk ceilings must specify the combined probability of cancer.

The problem of risk ceiling specification is somewhat complicated by the problem of voluntariness. Other contributors to this volume (see Chapters 10 and 13) have affirmed that people accept risk differently according to whether exposure is voluntary or not.[39-42] It therefore becomes necessary to consider that each type of exposure situation may have different quantitative risk ceilings so as to avoid selecting unacceptable ceilings in a given area. At the same time, the ceilings selected may not turn out to be in direct proportion to the measured perceptions of acceptability since more than this one factor must be taken into account in selecting them. For example, the need to avoid the risks inherent in excessive extrapolation of dose-response data might tend to cause certain ceilings to be adjusted upward.

Relating Risk Ceilings to a
Selected Goal

Ideally, in setting any goal for improvement it is desirable to know where we are now and where we should like to be, at least after a first step. In the case of industrially related cancer we have no firm knowledge now of where we are in relation to total cancer incidence. For this reason, it is suggested here that the first interim goal should be the following: to assure that industrially related cancer becomes *less* than a truly small fraction of today's total cancer incidence in the United States, regardless of what the industrial contribution actually is at the present time.

Since we do not know what the fractional contribution of total industrially related cancer incidence is, we also cannot know whether the fraction selected is above or below the present or future fractional contribution to total

incidence. If the fraction is below either of these unknown
values, then our efforts to achieve the goal are clearly
rightly directed. If the fraction is above either of these
unknown values, then the cancer burden related to industry
is already or will be only a small part of the total, and
because of the control philosophy of further subsequent
adjustment, will in time be made smaller yet.

Given the goal as a fraction, g, and the fraction, F,
of all people that contract cancer, the average individual
will have a risk of gF from industrially related cancer.
This average risk is related to risk ceilings, p_{ci}, arising
in different kinds of exposure situations through

$$gF = \sum_i p_{ci} f_i \qquad (1)$$

where f_i is the population fraction at risk in the ith kind
of exposure situation. The relation (1) assumes exposure
effects are additive: that is, the total risk from all situations can be obtained by adding the individual risks.
Eq. (1) by itself only defines average risk, however. For
a given individual who is exposed to all situations there
exists a maximum overall risk ceiling defined by a critical
probability

$$p_c = \sum_i p_{ci} \qquad (2)$$

which may be considerably larger than the average risk ceiling defined by Eq. (1), especially if some of the fractions
f_i are small. If positive synergistic effects between any
exposure situations should later be detected, Eqs. (1) and
(2) can be adjusted accordingly. Within exposure situations
more response functions can be added as need be. In terms
of the diagram of Figure 15.1, p_c may be identified with
the level C, dividing the region of unacceptable risk from
the region of lesser risks.

With Eqs. (1) and (2) together, the problem of risk
acceptance is reduced to specifying the risk ceilings to be
applied to different kinds of exposure situations. Though
straightforward in principle, this requires a series of
decisions involving the degree of voluntariness of exposure
and specification of the population fractions experiencing
the risk. However these decisions are made, they must satisfy Eq. (1) as an overall constraint.

As an illustrative example, suppose that g = 0.015 is
taken as an interim goal for the fraction of industrially

related cancer incidence. Suppose, further, that a small
number of broad kinds of exposure situations are ultimately
settled on, five involuntary and one (occupational exposure)
semi-voluntary; that of the five involuntary exposure situations so large a fraction of the population is at risk
that f_1 through f_5 may be set to unity with no important
error; and that an involuntary ceiling risk is judged to be
only one-tenth as acceptable as a semi-voluntary ceiling
after taking into account all the countervailing factors.
To calculate P_c it is only necessary to decide what f_6 is.
Those charged with this exercise could at this point decide
to estimate f_6, or, since the whole process is a societal
selection process, they could elect to include a small safety
factor and set $f_6 = 1$, knowing this would lead to somewhat
lower values of P_c. Taking the second choice, the equations
now yield a critical probability of $P_c = 3.75 \times 10^{-3}$, a ceiling
risk for semi-voluntary exposure of $P_{c6} = 2.5 \times 10^{-3}$, and a
ceiling risk for involuntary exposure of $P_{ci} = 2.5 \times 10^{-4}$,
with i = 1-5.

As in the woodcarving example, it is important to take
this kind of rough cut, at first, and to keep down the number
of kinds of exposure situations; later, if values of f_i can
be more closely defined, more kinds of exposure situations
can be defined and utilized.

Several matters must be kept in mind when evaluating
this concept: (a) Risk, as used here, is narrowly defined
as the probability that exposure in a given situation will
cause at least one adverse effect (e.g., malignancy) sometime in life after the exposure has occurred. (b) The goal,
when first selected, must be viewed as an interim goal,
subject to review after a suitable period has passed, thus
allowing learning in a stepwise process, and unless the P_c
selected should prove infeasible of attainment, review
should ordinarily not cause it to be raised. (c) New data
enabling the estimate of quantitative risks, when obtained,
should be used to review risk ceilings and to evaluate
relevant exposures. (d) If the method is to be used with
chemical carcinogens, as it must be, then animal data of
suitable quality must be relied on, since human epidemiological data take too long to collect and are too difficult
to interpret. Animal data, as already discussed, raise
their own questions, but no other data bearing a reasonable
relationship to human effects are as readily available. In
using animal data the most likely fit to the most reasonable
dose-response function should be utilized in describing
risk.[30] Moreover, it should be remembered that an exposure-
response function for humans that is based on animal data
is no more than a quantitative hypothesis.

Setting the Goal

Application of the process outlined here requires above
all the specification of the fraction, g. There is no
precise formula for doing this, but there are several major
guiding factors or criteria which may help in establishing a
rough range to be considered in making the selection. These
are of particular importance in the selection of the first
fraction since those making the selection will not yet have
the benefit of experience. Six guiding factors come to mind:

(1) Smallness. One obvious requirement is that the first
selected fraction specify a small part of total cancer incidence. Whether this requirement is met by setting a fraction
equivalent to a few percent or to less than one percent is
difficult to say. Comparing it to the 1-5% estimate of the
occupational contribution referred to earlier can offer some
guidance: a fraction equivalent to a percentage no higher
than one-half the maximum of this range could be considered
small. It would certainly be equivalent to a dramatic reduction in future incidence if compared with the higher 23-38%
estimate.

(2) Significance and detectability. The different
methods whereby estimates of industrially related cancer
incidence are obtained need to be examined to provide confidence limits. Ideally, the first goal fraction should not
only be small but also not significantly determinable at
normal confidence levels.

(3) Comparative risk. Wilson[43] has listed some occupational risks of death. Assuming a 30-year working career,
lifetime risks may be estimated from these figures. The
lowest risks of death so calculated lie in the range of
$2-3 \times 10^{-3}$. Manufacturing industry risk is found to be at
2.4×10^{-3}, and service industry risk at 2.7×10^{-3}. The highest
level given, for black lung disease in coal mining, leads to
a lifetime risk of 0.22. We have seen that an illustrative
fraction of 0.015 can yield, depending on assumptions, an
occupational risk of cancer incidence, not death, of 2.5×10^{-3}.
This risk would appear to correspond to the low end of the
spectrum of values presented by Wilson, and one might therefore question whether 2.5×10^{-3} is a true ceiling. It may
have to be somewhat higher to correspond to the level C in
Figure 15.1. Nevertheless, the value of risk comparisons in
selecting a goal expressed as a fraction is apparent. One
might reasonably select from the range of possible values
that satisfy the other guiding factors a value that lies
within the range of risks now apparently accepted but which,
for the hazard under consideration, represents a ceiling value.

(4) <u>Extrapolation</u>. Most quantitative animal data, when available at all, lie in a dose or exposure range yielding excess probabilities of cancer above 0.1. The sheer cost and difficulty of managing with reasonable confidence the large experiments needed for detecting lower responses make such experiments prohibitive for routine use. Only one study (the ED_{01} experiment) with excess cancer probability of 0.01 has been conducted,[44] and even this is well above the levels of interest in the realm of human health. It is therefore desirable, other things being equal, to set a ceiling as high as possible to avoid excessive extrapolation of high-dose animal data. "Unbalanced experiments" such as ED_{01} or as suggested by Farmer[45] may in the future improve the prospects for extrapolation, and experiments involving set mixtures, when an exposure situation involves more than one potential carcinogen, may cut down the degree of extrapolation required and therefore the uncertainty.

(5) <u>Feasibility</u>. Exploration of the technical and economic feasibilities of attaining different fractions, and of the attendant consequences, in a limited number of important, potential cases can offer guidance in the overall selection process.

(6) <u>Unacceptability/acceptability limit</u>. Finally, one concept needs to be borne in mind. The fraction to be selected should yield a critical probability that, for cancer, is the boundary between unacceptable risk and all other risks of various degrees of acceptability (See Figure 15.1). As such, the value sought is an upper limit or ceiling on risk, and conscious effort should be made to select a first fraction which, while complying with such guidelines as described above, is also compatible with the concept of a ceiling. Such a goal will result in a more rapidly implemented control of potential carcinogens and will facilitate the first rough step toward the ultimate objective.

How would the process of setting the fraction and defining the kinds and numbers of exposure situations be carried out in practice? It might become part of a national cancer prevention policy, which has been sought by the Interagency Regulatory Liaison Group[25] and the Regulatory Council.[26] Yet these kinds of organizations themselves cannot adopt the concept of unacceptable risk and its implementation through setting of a goal fraction. New legislation is needed, and a mechanism for setting the goal could well result from a legislatively mandated commission. Setting the goal fraction is a key societal decision[46] and would in effect involve a national goal. The review of the goal after some years, when its impact has become clear is, similarly, a societal action.

Application to Control or Regulation

Once the goal fraction and associated risk ceilings are set, it is possible to envision their application in control and regulation. Risk ceilings corresponding to a selected fraction can be used in specific exposure situations. Within a given kind of exposure situation, such as a specific water supply or a particular manufacturing unit, one may define two control regimes, depending on the ease with which control strategies can meet specified ceilings. Where it is easy to achieve risk levels below specified risk ceilings we shall speak of regime A and note that control strategies may well, but without serious socioeconomic disruption, set exposure levels below ceiling values down to levels of exposure where further reduction offers no further significant health bene-fits. Where achievement of stated ceiling risk levels is difficult in both a technical and socioeconomic sense, we shall speak of regime B. In this case full cost, risk-risk, and technical analyses should be made to find the control strategies that will lower the risk below the particular ceiling. Such strategies may well require special protective devices or clothing, partial restriction on manufacture, distribution, use or disposal, or even, at times, outright banning of an agent.

In both regimes, the net result of regulation when finally applied to all specific exposure situations and when followed up through monitoring and epidemiological studies to ensure that no significant additional risk is missed, should be to decrease risks of exposed individuals to less than the critical probability. As a result, the fraction of industrially related cancer incidence would fall below the national goal.

Regulation in regime A can give a new meaning to guide-lines that call for lowering exposures to the "lowest feasible level." Earlier we noted that in the abstract this prescription has no useful meaning. Bingham has stated that "to the extent feasible" means technological and economic feasibility.[47] This is a good start, but it needs to be defined even further. For regulation in regime A, in addi-tion to technological feasibility, operating and engineering means for reducing exposure should be technically known. Because regime A implies operation below risk ceilings, innovation should be reserved for future use. In addition, operating and technical solutions should not be such as to introduce new safety problems—otherwise they become

counterproductive in a regime that already lies below the
threshold of unacceptability.

From the standpoint of economic feasibility, costs
incurred should be either affordable without lessening of
competition (where this applies) or without harming the basic
ability to maintain the operability of the facilities, institutions, or businesses involved. In the case of business
enterprises, while the abilities to compete and grow may
decrease, they should not be eliminated in the name of feasibility. In short, regulation in regime A below the ceiling
should seek reduction of exposure where such reduction will
actually reduce risk, but with an understanding what feasibility takes full account of possible lost benefits, such
as employment, productivity, tax base, and useful products
and services.

It should be clear that application of the concept of
risk ceilings, and regulation in regimes A and B will involve
a continuing interlinked process of research and control
actions with the dual goal of (a) saving as many from contracting cancer as possible, and (b) providing for ongoing
risk reduction. It is important that these two goals be
approached concurrently.

Once a specific exposure situation containing potential
hazards is defined, a set of steps should involve estimating
and monitoring the extent of exposure; reviewing the materials present to determine whether any are known or potential
carcinogens and what is known about them; and carrying out
qualitative risk analyses and, where possible, quantitative
risk analyses with the purpose of evaluating possible human
hazards and risks. Concurrently, steps to reduce risk should
begin as active agents are so identified and analyzed as
described below.

To begin with, it is best to follow the concept of
regime A by taking five principal steps: (1) early reduction
of sources in number and volume by operating and engineering
modifications that can be made quickly; (2) determination of
what the constraints to further reduction are and which can
be most readily relieved to reduce further the most serious
exposures; (3) selection of feasible means of control under
regime A and definition of the resultant levels of exposure;
(4) development of the best estimate of dose-response functions for those substances present where possible; and (5)
determination from both qualitative and quantitative factors
whether the feasible control levels determined appear to
constitute an unacceptable risk or not. The last determination can only be made in a formal sense when necessary

dose-response data are on hand, but it should be considered in any event as the last step in control implementation. If the step leads to the conclusion that the risks are unacceptable, then work must proceed to see what the best actions are under regime B. In short, effective risk reduction starts with the assumption that we are in regime A, and continues to make progress along that path until information is developed that places the risk situation into regime B. Even when adequate quantitative data are available the uncertainties are such that deciding whether a risk lies in regime A or B will require the consideration of all relevant data and the exercise of judgment. One would hope that once quantitative information is adequate, it will be found that the risk is below the risk ceiling because of actions already taken.

Conclusion

The use of goal-setting, critical probability, and ceiling risk concepts addresses the real problem of risk control for materials such as carcinogens in which information is imperfect and there is a large region between clearly acceptable risks and clearly unacceptable risks. Together with a rational definition of feasibility, the concept of aiming toward a goal can allow rapid implementation of control and/or regulation on a broad front with minimal conflict. Moreover, the concept is well-suited to the use of performance rather than specification approaches to regulation, a distinction which makes for greater effectiveness, and it contains built-in incentives for private concerns to reduce risk as much as possible (e.g., maintenance of steady operation if ceilings are reduced on review). The ceiling risk concept in this sense also solves an ethical problem of regulation: it assures that people on the average can expect over time that control and regulatory actions taken would cause them to run no greater risk than the societally set ceiling risk for a given exposure situation, and no greater than a critical probability overall. Finally, this concept offers the possibility that if risk cannot be reduced to zero it can be aimed toward insignificance.

References and Notes

1. M. Eisenbud, *Environment, Technology and Health: Human Ecology in Perspective* (New York University Press, New York, 1978).

2. T.H. Maugh, "Chemical Carcinogens: The Scientific Basis for Regulation," *Science,* 201 (1978), 1200-1205.

3. New York (State), Governor's Panel to Review Scientific
    Studies and the Development of Public Policy on Pro-
    blems Resulting from Hazardous Wastes, *Report to the
    Hon. Hugh L. Carey, Governor of the State of New York,
    and Members of the New York State Legislature*, [L.
    Thomas et al.] (Albany, 8 October 1980).

4. D.E. Stevenson et al., "The Predictive Value or Other-
    wise of Conventional Carcinogenicity Studies," in
    Carcinogenicity Testing: Principles and Problems, ed.
    A.D. Dayan and R.W. Brimblecombe (University Park Press,
    Baltimore, 1978), p. 19.

5. T.H. Maugh, "Chemical Carcinogens: How Dangerous are
    Low Doses?," *Science*, 202 (1978), 37-41.

6. P.B. Hutt, "Quantitative Risk Assessment for Carcino-
    gens," *Legal Times of Washington*, 10 April 1979, p. 10.

7. J. Higginson, "Present Trends in Cancer Epidemiology,"
    Proceedings of the 8th Canadian Cancer Congress, Honey
    Harbour, Ontario (Pergamon, New York, 1969), pp. 40-45.

8. J. Higginson and C.S. Muir, "The Role of Epidemiology
    in Elucidating the Importance of Environmental Factors
    in Human Cancer," *Cancer Detection and Prevention*, 1
    (1976), 79-105.

9. J. Higginson, "A Hazardous Society? Individual versus
    Community Responsibility in Cancer Prevention," *Amer-
    ican Journal of Public Health*, 66 (1976), 359-366.

10. E.L. Wynder and G.B. Gori, "Contribution of the Environ-
    ment to Cancer Incidence: An Epidemiological Exercise,"
    Journal of the National Cancer Institute, 58 (1977),
    825.

11. R. Doll, "Strategy for Detection of Cancer Hazards to
    Man," *Nature*, 265 (1977) 589-596.

12. J. Higginson, "Special Lecture: Perspectives in Environ-
    mental Carcinogenesis," *The Cancer Bulletin*, 31, No. 3
    (1979), 50.

13. J. Higginson, "Cancer and Environment: Higginson Speaks
    Out," *Science*, 205 (1979), 1363.

14. P. Cole, "Cancer and Occupation: Status and Needs of
    Epidemiologic Research," *Cancer*, 39 (April supplement,
    1977), 1788-1791.

15. K. Bridbord and other contributors,"Estimates of the Fraction of Cancer in the United States Related to Occupational Factors," Prepared by: National Cancer Institute, National Institute of Environmental Health Sciences, National Institute for Occupational Safety and Health, Washington, (15 September 1978).

16. Editorial, *Lancet*, 2, No. 8102 (1978), 1238-1240.

17. W.R. Barclay, "Science Reporting Alarm to the Public" (Editorial), *JAMA* 242 no. 8 (24/31 August 1979), 754.

18. American Industrial Health Council (1075 Central Park Avenue, Scarsdale, New York 10583), *A Reply to: "Estimates of the Fraction of Cancer Attributable to Occupational Factors,"* [Ref. 15] (AIHC, Scarsdale, New York, 23 October 1978).

19. J. Higginson and C.S. Muir, "Environmental Carcinogenesis: Misconceptions and Limitations to Cancer Control" (Guest editorial), *Journal of the National Cancer Institute*, 63 no. 6 (December, 1979), 1291.

20. E. Whelan, *Cancer in the United States: Is There an Epidemic?* (American Council on Science and Health, New York, 1978).

21. J.E. Enstrom, "Cancer Mortality Among Low-Risk Populations," *CA-A Cancer Journal for Clinicians*, 29, No. 6 (November/December, 1979), 352.

22. J.G. Martin (Congressman), "The Chemical Nature of Politics (and Vice Versa)," *The Chemist*, 55 (May, 1978) 5.

23. W.C. Wampler (Congressman), "Remarks Before the Third Annual Conference on Pesticide Regulation and Registration," Washington, 19 October 1979.

24. Occupational Safety and Health Administration, "Identification, Classification and Regulation of Toxic Substances Posing a Potential Occupational Carcinogenic Risk, Proposed Rule and Notice of Hearing" *Federal Register*, 42 (1977), 54149.

25. Interagency Regulatory Liaison Group, "Scientific Bases for Identification of Potential Carcinogens and Estimation of Risk," *Federal Register*, 44 (1979), 39858.

26. Regulatory Council, "Regulation of Chemical Carcino-
     gens," *Federal Register*, 44, (1979), 60038.

27. Environmental Protection Agency, "National Emission
     Standards for Hazardous Air Pollutants: Proposed Pol-
     icy and Procedures for Identifying, Assessing, and Reg-
     ulating Airborne Substances Posing A Risk of Cancer,"
     Federal Register, 44 (1979), 58642.

28. L.J. Carter, "How to Assess Cancer Risks," *Science*,
     204 (1979), 811-816.

29. Occupational Safety and Health Administration, "Iden-
     tification, Classification and Regulation of Potential
     Occupational Carcinogens" *Federal Register*, 45 (1980)
     5001.

30. D.R. Calkins et al., *Identification, Characterization,
     and Control of Potential Human Carcinogens: a Frame-
     work for Federal Decision-Making; A Staff Paper...*
     (Office of Science and Technology Policy, Washington,
     1 February 1979).

31. William W. Lowrance, *Of Acceptable Risk: Science and
     the Determination of Safety* (William Kaufmann, Los
     Altos, California, 1976).

32. Food and Drug Administration, "Chemical Compounds in
     Food-Producing Animals," *Federal Register*, 44 (1979),
     17070.

33. T.C. Byerly, "USDA Policy on Carcinogens" (Contract
     43-32R7-9-1100), 2 July 1979; D.R. Calkins et al.,
     Ref. 30.

34. Environmental Protection Agency, "Request for Comments
     on Water Quality Criteria for 27 Toxic Water Pollu-
     tants," *Federal Register*, 44 (1979), 15926.

35. Bill Richards, "OSHA Plans to Name Hundreds of Suspected
     Carcinogens," *Washington Post*, 20 November 1979.

36. A.C. Upton, Statement on the National Cancer Program
     Before the Subcommittee on Health and Scientific
     Research, Senate Committee on Human Resources, Wash-
     ington, 7 March 1979.

37. J.F. Young, Written Comments by the General Electric
     Company to the Regulatory Council on the Regulatory

Council Statement on Regulation of Chemical Carcinogens (Ref. 26), 13 November 1979.

38. Food Safety Council, "Principles and Processes for Making Food Safety Decisions," *Food Technology,* 34 (March, 1980), 1-45.

39. B. Fischhoff et al., "How Safe is Safe Enough? A Psychometric Study of Attitudes Towards Technological Risk and Benefits," *Policy Sciences,* 8 (1978), 127-152.

40. C. Starr, "Social Benefit Versus Technological Risk," *Science,* 165 (1969), 1232-1238.

41. W.D. Rowe, "Governmental Regulation of Societal Risks," *The George Washington Law Review,* 45, No. 5 (August, 1977), 944-968.

42. W.D. Rowe, "Risk Assessment Methods and Approaches" in *Society, Technology and Risk Assessment,* ed. Jobst Conrad (Academic Press, London, 1980), pp. 3-29.

43. R. Wilson, Direct Testimony Before the United States Department of Labor Assistant Secretary of Labor for Occupational Safety and Health Administration, OSHA Docket No. H-090, Washington, 1978.

44. J.A. Staffa and M.A. Mehlman, Editors, et al., "Innovations in Cancer Risk Assessment (ED_{01} Study) *Journal of Environmental Pathology and Toxicology,* Special Issue 3, No. 3 (1980).

45. J.H. Farmer et al., "An Unbalanced Experimental Design for Dose Response Studies," *Journal of Environmental Pathology and Toxicology,* 1 (1977), 293.

46. D. Bazelon, "Risk and Responsibility," Editor's Page, *Chemical and Engineering News,* 57 (10 September 1979), 5.

47. L.R. Ember, "OSHA's Bingham Still A Happy Warrior," *Chemical and Engineering News,* 58 (18 February 1980), 23.

Richard C. Schwing

16. Longevity Benefits and Costs of Reducing Various Risks

> *That which is purely practical, containing no*
> *element of idealism, may sustain existence and*
> *to that extent be valuable, but it does not*
> *civilize.*[1]

Many actions by institutions, both public and private, are designed to reduce or eliminate risk. However, although the costs of all risk elimination must be borne by a relatively fixed budget of scarce resources and are comparable, the benefits of various programs and policies are difficult to compare.

The comparisons are hindered by the problems in measuring social benefits. Because matters of life and death are loaded with emotional and subjective overtones, policy decisions that impinge on these matters are burdened with many attributes that are difficult to quantify.

Furthermore, even the quantified benefits derived by reducing various risks have little meaning to the average person when expressed as probabilities or rate changes. The personal significance, if any, of reducing the probability of being killed in a 5-mile trip from 0.0000002 to 0.0000001, aside from one figure being half the other, is very nebulous.

In a global context, "lives saved" is also a very misleading measure of a policy. What does it mean to save a life when we are all mortal? It is more accurate to claim that a given action removes or reduces a specific category of mortality that is in competition with other categories of mortality. As death is certain, there is no risk (i.e.,

Adapted by permission from Technological Forecasting and Social Change, vol. 13 (May 1979), pp. 333-345.

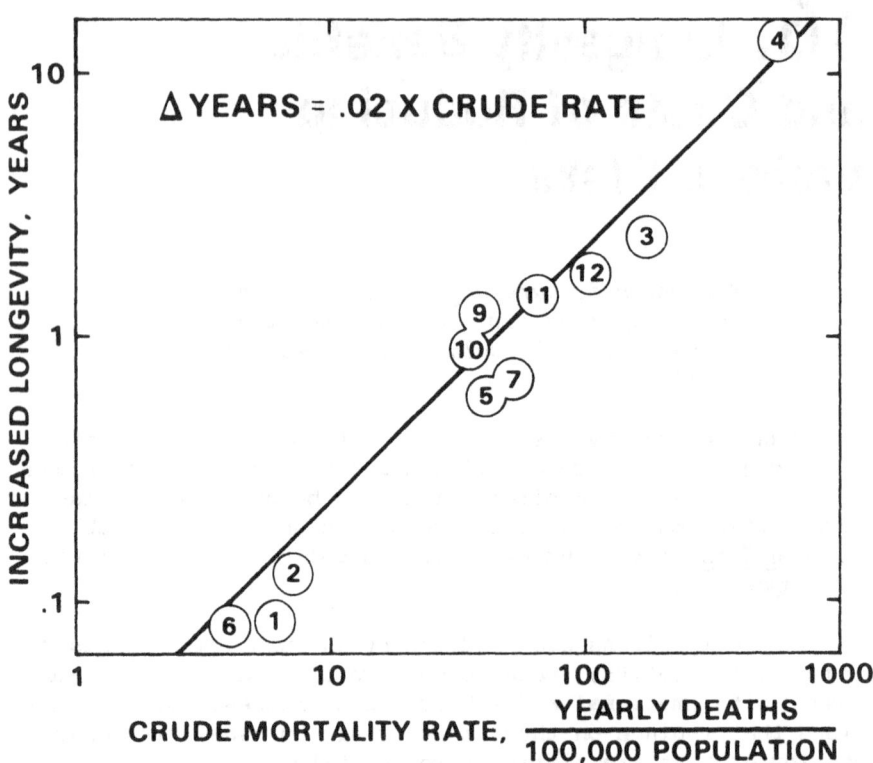

Fig. 16.1. Longevity gain due to elimination of individual disease categories for U.S. males, based on the data of Preston et al. (ref. 2). Numbered symbols in the figure are defined as follows: (1) respiratory tuberculosis; (2) other infections; (3) malignant and benign neoplasms; (4) cardio-vascular disease; (5) influenza, pneumonia, bronchitis; (6) diarrhea, gastritis, enteritis; (7) certain degenerative diseases; (8) complications of pregnancy (not applicable); (9) certain diseases of infancy; (10) motor vehicle accidents; (11) other accidents and violence; (12) all other and unknown causes.

uncertainty) imposed on our mortality.   In other words, death
does not involve chance; it is certain.   It is the timing
that is uncertain.   Before we die, however, many activities,
both voluntary and involuntary, present a risk of premature
death.   It is the word "premature" that gives us a hint of
a helpful metric.   The extent to which one dies before one's
time, then, may be more meaningful than mortality rates.

To deal with these issues, this chapter focuses on
longevity changes as a substitute for mortality risk or
"lives saved."   (In contrast to other chapters in this volume, we use the word "risk" narrowly to denote the expected
mortality rate.)   Our purpose is to provide an interpretation
of mortality risk that is intuitively accessible to the non-
expert.   To achieve this end, we must by necessity begin with
a somewhat technical discussion of mortality statistics.

The Relation Between Mortality and Longevity

Quantitative Correlations

Life Tables for National Populations[2] provides a data
base that allows a study of a wide range of risks.   Tabulations of data accumulated since 1900 are available for males
and females of several populations for 12 categories of
mortality risk.   The tabulations are based on statistics for
the respective countries.   In these tables the results presented are calculations, based on the mortality rates, that
estimate the number of persons surviving to age X if specified causes were eliminated.   In addition, added years of
life and added years of work are estimated if specified
causes could be eliminated.   For example, if all cancers
were eliminated, it is estimated that the average male in
the United States would gain 2.265 years of longevity.   The
calculations are made assuming that the tolls due to the
remaining 11 risk categories are independent of the cause
eliminated in each case.

One might expect the longevity change to be a function
of the mortality rate, the age distribution of the impact,
and the age distribution of the exposed population.   To test
this, added years of life (calculated from 1964 mortality
data[2]) for U.S. males for the elimination of each category
is plotted as a function of the crude mortality rate in
Fig. 16.1 A reasonably linear relationship exists over
three orders of magnitude.   Some scatter arises because
different categories impact on different age groups in the
population.   Even the extremes, however, such as the category affecting the elderly, cardiovascular, and the category for infancy, fall reasonably close to the straight

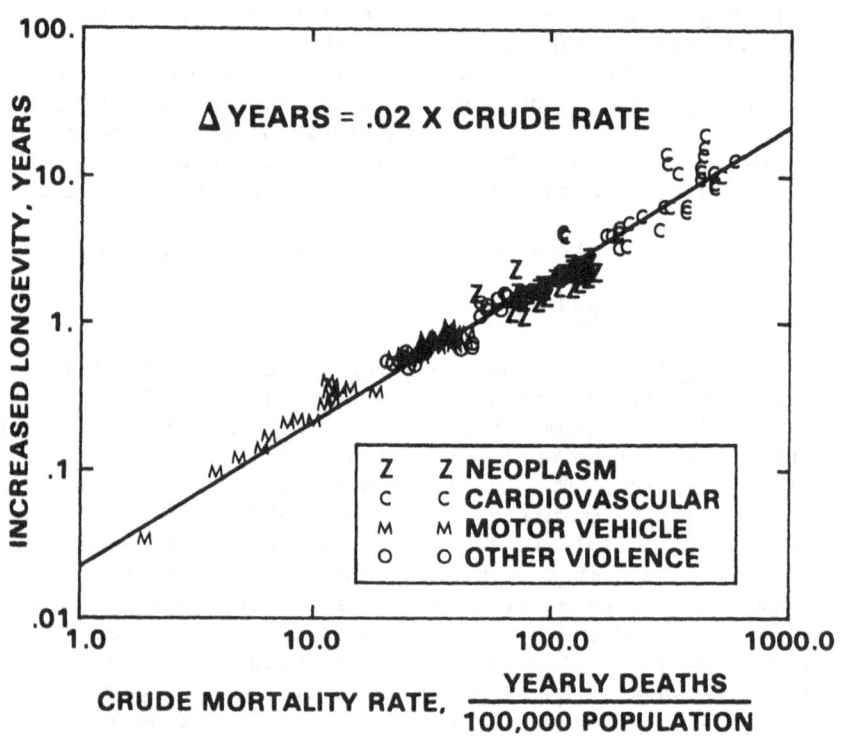

Fig. 16.2. Increased longevity due to elimination of four disease categories, for both males and females and several different years. Data for the United States, Australia, and Canada.

line having the equation

$$\Delta \text{ years} = 0.02 \times \text{crude rate},$$

where the crude rate represents the annual number of deaths in a population of 100,000. No age adjustments are imposed.

Note that categories designated 5,7,9,10, and 11, in Fig. 16.1, which all cluster near the center of the range, each provide approximately one year of added longevity if the category could be completely eliminated.

To test the generality of the correlation, the same variables from three populations, the United States, Australia, and Canada, are plotted in Fig. 16.2 for four categories, neoplasms (cancers), cardiovascular fatalities, violence, and motor-vehicle accidents, for both males and females and several different years. Again the equation given above provides a reasonable representation of the data for our exploratory purposes. The scatter envelope, which is visually reduced with a log-log plot, is within 40% of the line. Linearity probably results from the fact that small changes in longevity, which is a nonlinear function of the mortality rate, can still be approximated by the first term in a Taylor series if the change in mortality rate is sufficiently small.

Since the correlation is reasonably general it is possible to estimate longevity change for mortality rates which are not presented in the life tables[2]. It is also possible to estimate potential longevity change derivable from programs designed to reduce or eliminate various categories of risk.

Conversion of Various Death Rates to Longevity

The correlation line derived in the previous section may be used to place in perspective a variety of other mortality risks. To this end, U.S. death rates for the general population are placed on the previously derived correlation line as shown in Fig. 16.3. The labels for the general population rates are placed above the correlation line in Fig. 16.3. It should be emphasized that the corresponding longevity gains derived by theoretically eliminating the respective death-rate categories are valid only insofar as the correlation line established in Figs. 16.1 and 16.2 is valid.

Risk rates that apply to specific groups, having higher or lower than average exposure rates, are also presented in

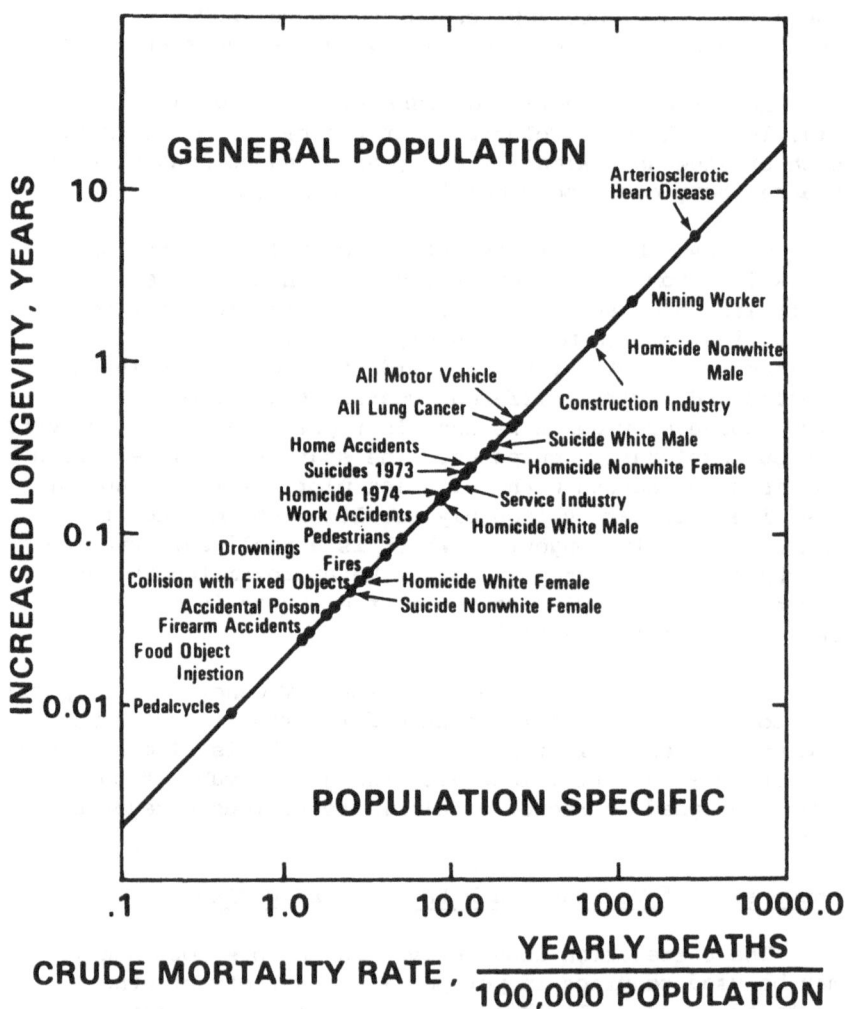

Fig. 16.3. Increased longevity vs. crude mortality rate for selected risks. To obtain the plot shown, we began with the straight line derived from Figs. 16.1 and 16.2 and then located mortality rates on this line. Data for the general population are labeled above the line, data for specific populations below the line.

Fig. 16.3. These rates are labeled below the line. The longevity gains again depend on the validity of the correlation and on the assumption that the subgroup has a mortality rate similar to the general population.

Only if one has an interest in a specific comparison does the tabulation provide particular insight. I might point out, however, that our perception of risk does not always compare to the magnitude of the risk. For example, only recently have homicide rates (10.3 deaths per year per 100,000 persons in 1974) approached the magnitude of suicide rates for the general population (12.3 deaths per year per 100,000 persons in 1973), and the daily ingestion of food objects is as risky as occasional firearm mishandling (1.2 versus 1.3 deaths per year per 100,000 persons, respectively). Furthermore, the risks to specific narrowly defined subgroups can rival the arteriosclerotic heart-disease rate, which is the highest category in the general population. The possible reasons for varying perceptions of equivalent mortality risks are discussed elsewhere in this volume; for example, Chapters 8 and 10.

Comparison of Measures to Reduce Risk

Measures to reduce risk, often in the form of legislative action, are not always imposed with perspective with regard to the benefits to be achieved. With respect to industry, legislation is imposed to reduce the mortality due to safety hazards and air pollution. The mortality and hence longevity benefits of actions can be estimated by using published estimates of the effectiveness of various actions to calculate the changes in mortality rates.

Two air-pollution values are presented in Fig. 16.4 that are based on elasticity estimates from regression studies.[6,7] The elasticities represent the estimated change in the total mortality rate due to a change in air-pollution level. Although elasticities are considered valid only near the mean value of the air-pollution level, we have extrapolated to zero pollution levels. Extrapolation to zero levels should be viewed with great caution. At the present time, however, these calculated net changes due to total pollution reduction represent the only estimates for the benefits of abatement. The longevity benefits can be read from Fig. 16.4. The estimates are presented here solely to illustrate a speculated gain due to a policy action. The quantitative support for these estimates is extremely weak and furthermore is based on statistical methods that cannot be construed as showing cause and effect.

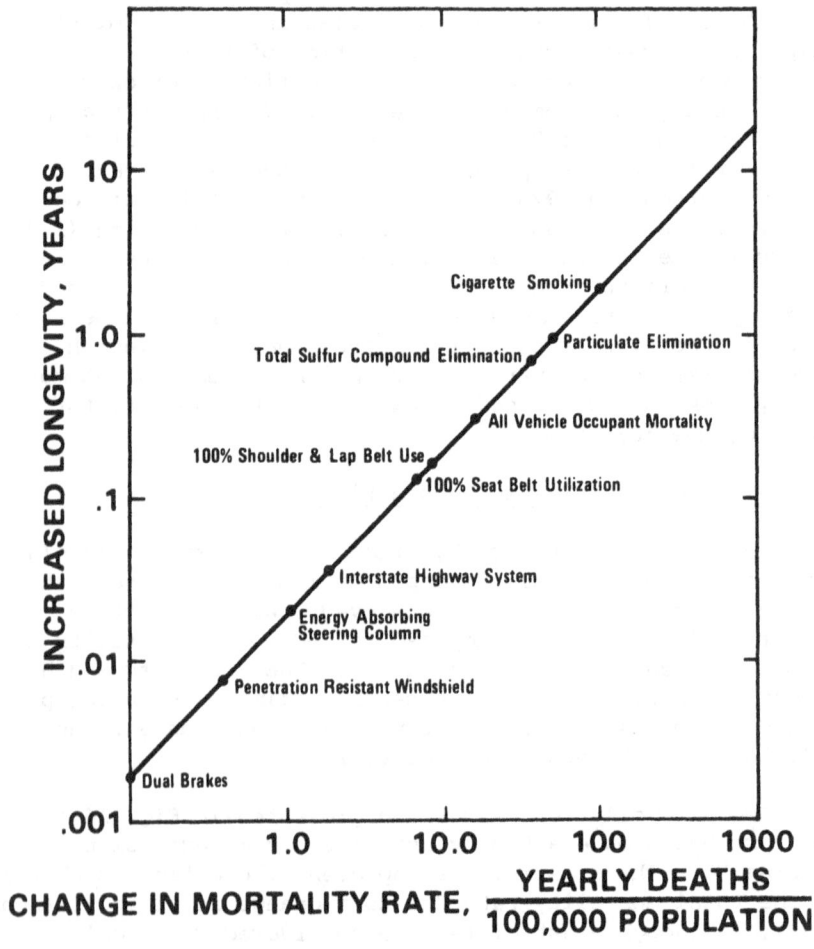

Fig. 16.4. Estimated increase in longevity vs. mortality for several air pollution and accident abatement programs. The mortality data were obtained from the literature, and as in the case of Fig. 16.3, were plotted on the straight line derived via Fig. 16.1 and 16.2.

The benefits of vehicle safety devices and programs[8-12] are also presented in Fig. 16.4. The estimates were made assuming 100% use factors and installation rates. They do not, however, assume 100% effectiveness. The values used represent estimates of effectiveness tabulated by Peltzman.[13] Peltzman considered the values optimistic and his study provides some evidence that these devices and programs have not in fact provided the gains tabulated in Fig. 16.4. Again, the data are presented as illustrative only.

To compare alternatives, the policymaker can now organize the benefits to be derived from a variety of expenditures on such a graph, decide where the greatest gains can be achieved per dollar expended, and efficiently allocate the finite resources available in the society.

By viewing various risks in the context of other risks presented in Fig. 16.4, one gains perspective in assessing the magnitude of the risks and the benefits that might be realized by their elimination. The estimated benefits range from 0.002 years' longevity for dual brakes to 2 years for the elimination of cigarette smoking, thus differing by a factor of 10,000.

Caveats

The reader should be cautioned on the use of these data in drawing general conclusions concerning benefits. Uncertainties in the accuracy of all effectiveness measures and elasticities should always be emphasized. Furthermore, only one measure or attribute of the benefits (i.e., longevity) has been proposed in this study. A host of attributes must also enter into decisions on policy. These attributes range from timelessness and pain and suffering to medical care resources and distributive justice, all aspects of the quality of life. In addition, the quality of the extended longevity as raised by Zeckhauser[14] becomes extremely important as the accuracy of the impact assessments improves.

Expenditures to Reduce Risk and Increase Longevity

Equally important to the policymaker is a tabulation of costs to achieve these ends. With cost information, cost/benefit or cost/effectiveness measures can then be used as an aid to establish priorities in making policy decisions.[15]

In any real society having limited resources, not all desired programs to save lives or extend longevity can be implemented. Some remain unfunded. Hopefully, those pro-

grams having the greatest efficiency in achieving desired
goals will be implemented before the less efficient ones.
If not, not only can the cost of less efficient programs be
expressed in terms of dollars, but the cost can also be
expressed in terms of alternative lives (or other social
values) sacrificed.

To gain perspective in comparing programs having common
goals, the costs and benefits of a number of medical, environmental, and safety programs from a variety of sources are
compared. Since the material has been derived from a variety
of studies in several sectors utilizing different methods,
the accuracy of some numbers can be questioned. Our purpose
is to propose this perspective as an aid in policy decisions;
it is not to supply a data base for decision-making since
high uncertainties remain.

Applicable Literature and Method

Estimates of program costs and program effectiveness
in terms of either longevity gain or reduced mortality have
recently been published by a number of authors. Each of the
studies is described in the following paragraphs.

Longevity gains and the corresponding dollar value per
person year of longevity were derived from each study or
used directly as published. They are presented on a common
1975 dollar basis for comparison. The impact was calculated
using the relationship

$$\Delta \text{ years} = 0.02 \times \text{crude mortality rate.}$$

Where necessary, a 20-year longevity gain per life saved was
used.

**Health, Education, and Welfare Analysis of Cancer-
Control Programs (1972):**[16] The U.S. Department of Health,
Education, and Welfare analyzed the cost/effectiveness of
various cancer programs and compared them to other life-
saving programs. This cost/effectiveness analysis served as
a useful aid to the government in framing pertinent questions and in improving the chances of moving in directions
of social improvement. R. N. Grosse, author of this paper,
argues that "the truly moral problem is not to distinguish
between good and evil, but rather to select appropriately
among alternative goods." Tabulated estimates by Grosse for
the cost per death averted for breast, head and neck, and
colon-rectum cancer programs are used. Graphical data by
Grosse provide the basis for the values for deaths averted
by lung cancer, syphilis, and tuberculosis programs. Bene-

fits from vehicle safety by Grosse are not presented, as his analysis for this category was cursory.

Benefit/Cost Analysis of Traffic-Safety Programs (1973):[17] V. J. Niklas of West Germany tabulated cost/benefit ratios and reductions in vehicle deaths and injuries, as well as program costs for a variety of highway-safety programs. The ratios of his tabulated costs to deaths averted are used for expressway lighting, truck under-ride protection, head rests, shoulder and lap belts, highway rescue helicopters, and highway rescue cars. To scale the data to the United States, the ratio of U.S. motor-vehicle fatalities to those of West Germany in 1969 was used (i.e., 56,400/16,464 = 3.38). The numbers are presented in terms of longevity benefits only. No account is taken of injury or property-loss reductions.

Passive and Active Restraint Systems: Performance and Benefit/Cost Comparison (1975):[18] L. M. Patrick of Wayne State University has provided a most complete analysis of passenger-restraint systems. Patrick's data, presented as the 10th-year cost per fatality eliminated, are used. Active and passive systems including air-bags, harnesses, and lap belts as well as passive torso belt and knee-bar systems, are analyzed. Patrick ranks the alternatives in terms of total fatality reduction and cost per life saved. He concludes that the mandatory harness is superior to all other systems in all comparisons, with approximately 100,000 lives saved over the first 10 years, about twice as many as would be saved by the other systems.

Kidney Dialysis and Transplant Cost/Effectiveness (1968):[19] Most programs designed to save lives deal with unidentified or "statistical" victims. On the basis of statistical data, an estimate of the number of victims can be made and another estimate of the program effectiveness calculated. In contrast, the study of kidney dialysis and transplant effectiveness assumes the victim has been identified and that only the victims of kidney disease are involved in the treatment decision. The "willingness to pay" for such a program increases drastically when the victim's identity is known.[20]

The data presented by Klarman[19] are based on existing programs and are included in this compilation to compare societal behavior for the identified victim to that corresponding to the unidentified "statistical" victim. It is interesting to note that the results of this study, based on actual mortality changes for kidney disease victims with and without dialysis or transplant, are presented directly

as cost per year of life rather than cost per "life saved."
The data presented also reflect current costs for home and
hospital treatment.[21]

Public Programs to Prevent Heart-Attack Fatalities
(1973):[22] The Rand Corporation in Santa Monica, California,
sponsored a program to evaluate health programs that save
lives by structuring the problem as one of decision-making
under uncertainty and by developing a willingness-to-pay measure of the worth of the program. The study, by J. P. Acton,
considered reducing coronary deaths by an ambulance program,
a mobile coronary care system, and a pretreatment screening
program. Although there exists no assessment of the actual
cost of existing coronary care programs, a recently instituted volunteer program in Seattle for coronary victim
treatment appears to be nearly as effective as the Rand
estimate.[23] The Rand study also determined people's expressed value of life saving as well as the expected program cost
per life saved. Two methodologies, a measure of the value
of livelihood and a measure of willingness to pay, provide
values of $32,000/life and $33,600/life, respectively, or
approximately $1,600 per year of longevity.

Air-Pollution Abatement Estimates (1976):[23] In a
cost/benefit study of automobile-emission reductions,[24]
information was presented which allows one to estimate the
costs and benefits of various programs legislated by the
Clean Air Act. By using the highest statistical associations available from regression studies that associate mor-
tality-rate changes with air pollutants and making the
assumption that oxidant associations are equal to nitrogen
compounds in magnitude as well as considering all nitrogen
compounds to be derived from NO_x emissions, it is possible
to associate a longevity benefit with the abatement of auto
pollution to 1977-1979 standards (1.5 g/mile hydrocarbon,
15 g/mile carbon monoxide, and 2.0 g/mile nitrogen oxide).

It should be emphasized that the longevity benefits
for automotive pollutants have not been reproduced by other
epidemiological studies. These benefits are considered
upper-bound numbers and are presented here as illustrative
of a lower bound (i.e., a low value for $ per person year
of longevity) in the effectiveness measure for abating
these pollutants.

Thus far, total benefits and total costs have been used
to derive impact and effectiveness measures. When adequate
data are available, it is more appropriate to consider the
marginal effectiveness derived from a marginal cost increase.
Environmental Protection Agency (EPA) estimates for benefits

due to additional CO abatement can be used to illustrate a marginal analysis. In the "300-day study" entitled <u>Air Quality, Noise, and Health Panel Report to Task Force for Motor Vehicle Goals Beyond 1980</u>,[25] reduced cardiac mortality benefits are tabulated for a 20-year period. The EPA estimates that two lives will be "saved" in the nation in 20 years by going from a 15 g/mile to a 3.4 g/mile CO standard. The first decade of cost of achieving this improvement is $1.0 billion, based on the methods in Schwing et al.[24] Therefore, we should be willing to spend at least $500 million per life or $25 million per person year of longevity to justify this standard, which is traditionally justified on the basis of either mortality or morbidity risks.

Similar data for sulfur compounds[24] allow an estimate of a value of longevity for sulfur removal from stationary power stacks. We use the EPA abatement-cost estimate of $0.003 per kWh to estimate the cost of the sulfur scrubbing technology. The corresponding longevity change for the U.S. population is 0.46 years for sulfur controls. In the case of sulfur abatement, a number of epidemiology studies confirm the order of magnitude of these benefits.

Comparison of Effectiveness Measures

Effectiveness measures for these 28 programs are presented in Fig. 16.5. The most striking characteristic of the data representing actual programs or programs under serious consideration is the fact that cost/effectiveness estimates range over five orders of magnitude and the impacts range over nearly six orders of magnitude. The <u>most</u> effective measure is the coronary care ambulance system located in the lower right-hand corner of Fig. 16.5. The <u>least</u> effective measure having the <u>lowest</u> impact, because only two lives are saved, is the marginal emission level change for carbon monoxide from 15g to 3.4 grams per mile.

Department of Transportation Data

A similar ranking of programs was published in the <u>National Highway Safety Needs Report</u>.[26] Under the leadership of W. T. Coleman, Jr., the Department of Transportation tabulated estimates of the cost/effectiveness of various expenditures on highway safety in $/fatality. Coleman's report to Congress contains a complete description of the expenditure computation along with the appropriate caveats. Since the data included the number of fatalities forestalled and total implementation costs, it was possible to present the data in an impact and longevity context as in Fig. 16.5. The DOT information is presented in Fig. 16.6 as an internally consistent set of data. Of the safety measures

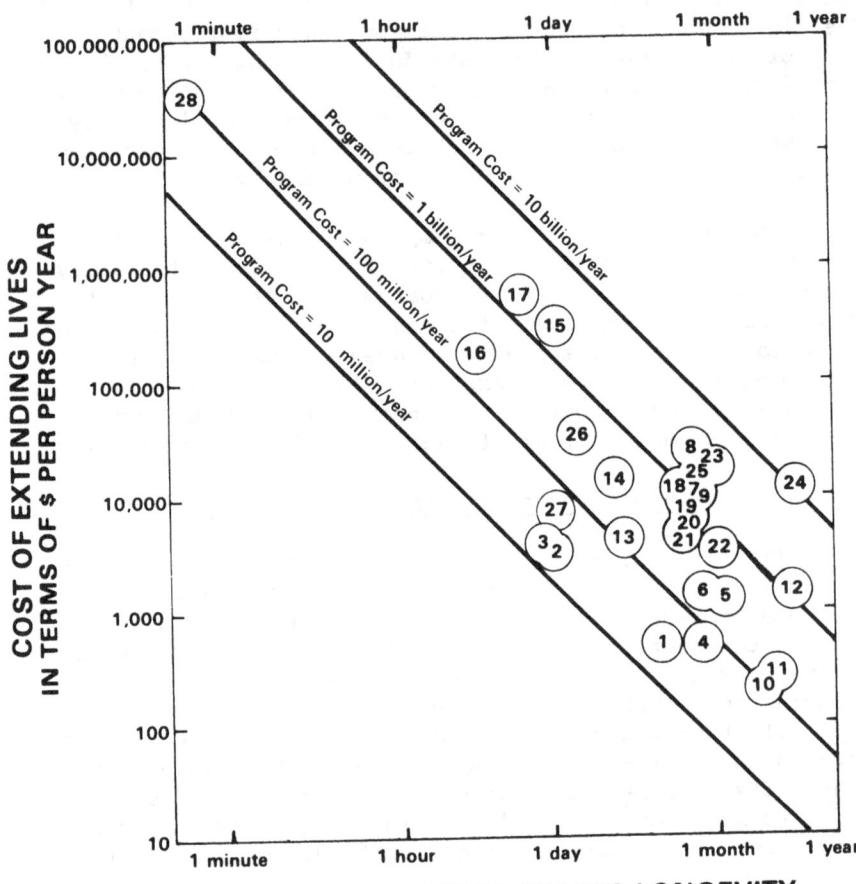

Fig. 16.5. Cost effectiveness of risk abatement, as measured in dollars per year of increased longevity. The 28 programs plotted in the graph are identified by number in Table 16.1.

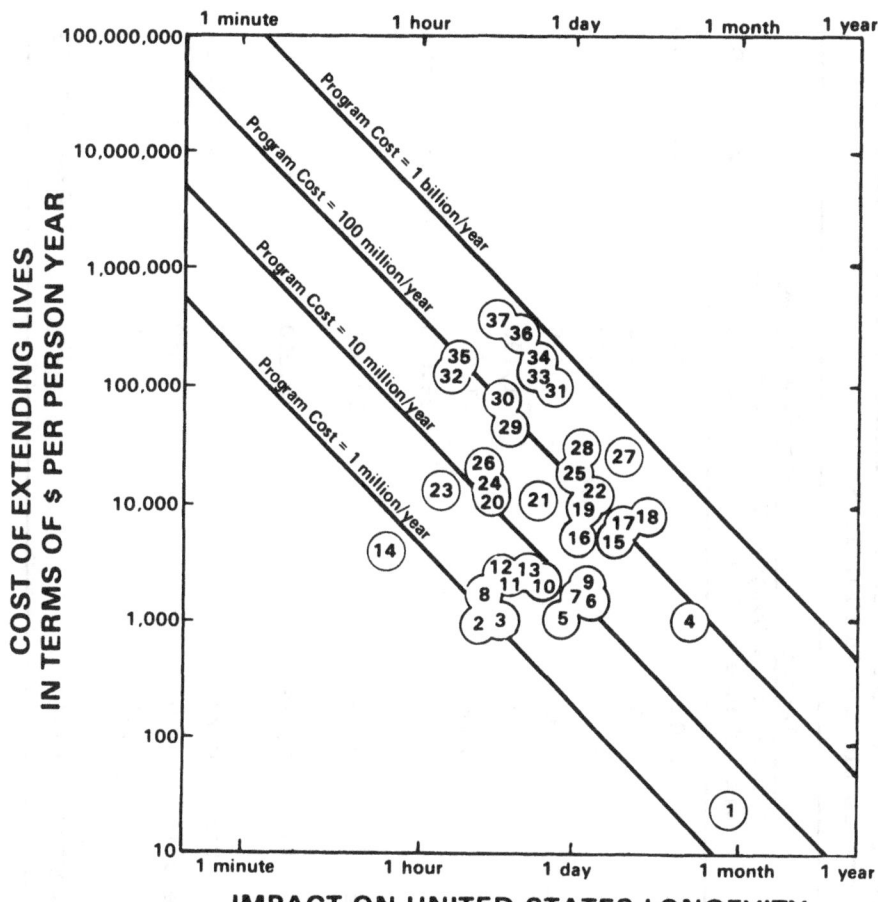

<u>Fig. 16.6.</u> Cost effectiveness of 37 highway safety measures expressed in dollars per year increased longevity. All of the data have been converted from a Department of Transportation survey described in ref. 26. The measures plotted in the figure are identified by number in Table 16.2.

Table 16.1. Summary of programs for which cost-effectiveness has been estimated by the author. Numbering corresponds to symbols plotted in Fig. 16.5.

	Program Description	Impact Population's Longevity Change in Yrs.	Cost-effectiveness $ Per Person year of Longevity	Source of Data	Reference
1.	Breast cancer	0.027	480	Dept.of H.E.W.	2
2.	Head and neck cancer	0.0028	3,475	"	"
3.	Colon-rectum cancer	0.0027	3,670	"	"
4.	Lung cancer	0.065	475	"	"
5.	Syphilis	0.102	1,190	"	"
6.	Tuberculosis	0.065	1,350	"	"
7.	Kidney transplant (victim identified)	0.056	25,000	Johns Hopkins U.	19
8.	Home kidney dialysis (victim identified)	0.056	10,000	"	21
9.	Hospital kidney dialysis (victim identified)	0.056	10,000	"	19
10.	Coronary ambulance system	0.22	192	Rand Corp.	22
11.	Mobile coronary care unit	0.30	258	"	"
12.	High-risk coronary screening	0.42	1,440	"	"
13.	Highway rescue helicopters	0.013	4,180	West Germany	17

Table 16.1, continued

Program Description	Impact Population's longevity Change in Yrs.	Cost-effectiveness $ Per Person year of Longevity	Source of Data	Reference
14. Highway rescue cars	0.011	14,390	West Germany	17
15. Lighting of all expressways	0.0033	310,000	"	"
16. Truck under-ride protection	0.00063	180,000	"	"
17. Four headrests per car	0.0016	580,000	"	"
18. Air bag	0.043	11,000	Wayne State U.	18
19. Air bag + 20% lap belt	0.049	9,500	"	"
20. Passive harness	0.051	6,500	"	"
21. Torso belt and knee bar	0.043	4,400	"	"
22. Mandatory harness	0.092	3,250	"	"
23. Four lap and shoulder belts per car 70% usage	0.085	17,100	West Germany	"
24. Sulfur scrubbing $0.003/kwh	0.46	11,300	GM, EPA	24,25
25. 1.5 gm/mile HC auto std.	0.061	15,400	"	"
26. 15 gm/mile CO auto std.	0.005	35,000	"	"
27. 2.0 gm/mile NO_x auto std.	0.035	1,280	"	"
28. 3.4 gm/mile CO auto std.	0.000001	27,500,000	"	"

Table 16.2. Cost-effectiveness of 37 highway safety measures
expressed in dollars per life saved. The data were compiled
by the U.S. Department of Transportation (ref. 26) and were
used by the author to obtain Fig. 16.6. The numbering of
measures corresponds to the numbering of symbols in Fig. 16.6.

	Highway Safety Measure	Number of Fatalaties Forestalled	Dollars per Life Saved
1.	Mandatory safety belt usage	89,000	500
2.	Highway construction and maintenance practices	569	20,000
3.	Upgrade bicycle and pedestrian safety curriculum offerings	649	20,000
4.	Nationwide 55 mph speed limit	31,900	21,000
5.	Driver improvement schools	2,470	21,000
6.	Regulatory and warning signs	3,670	34,000
7.	Guardrail	3,160	34,000
8.	Pedestrian safety information education	490	36,000
9.	Skid resistance	3,740	42,000
10.	Bridge rails and parapets	1,520	46,000
11.	Wrong-way entry avoidance techniques	779	49,000
12.	Driver improvement schools for young offenders	692	52,000
13.	Motorcycle rider safety helmets	1,150	53,000
14.	Motorcycle lights-on practice	65	80,000
15.	Impact absorbing roadside safety devices	6,780	108,000
16.	Breakaway sign and lighting supports	3,250	116,000
17.	Selective traffic enforcement	7,560	133,000
18.	Combined alcohol safety action countermeasures	13,000	164,000
19.	Citizen assistance of crash victims	3,750	209,000

Table 16.2, continued

	Highway Safety Measure	Number of Fatalaties Forestalled	Dollars per Life Saved
20.	Median barriers	529	228,000
21.	Pedestrian and bicycle visibility enhancement	1,440	230,000
22.	Tire and braking system safety critical inspection--selective	4,591	251,000
23.	Warning letters to problem drivers	192	263,000
24.	Clear roadside recovery area	533	284,000
25.	Upgrade education and training for beginning drivers	3,050	385,000
26.	Intersection sight distance	468	420,000
27.	Combined emergency medical countermeasures	8,000	538,000
28.	Upgrade traffic signals and systems	3,400	610,000
29.	Roadway lighting	759	936,000
30.	Traffic channelization	645	1,680,000
31.	Periodic motor vehicle inspection--current practice	1,840	2,120,000
32.	Pavement markings and delineators	237	2,700,000
33.	Selective access control for safety	1,300	2,910,000
34.	Bridge widening	1,330	3,460,000
35.	Railroad-highway grade crossing protection (automatic gates excluded)	276	3,530,000
36.	Paved or stabilized shoulders	928	5,800,000
37.	Roadway alignment and gradient	590	7,680,000

considered, the cost/effectiveness measures range from
$25.30 per person year of longevity to $384,000 per person
year of longevity.

Since some of the programs such as air-bag and harness
systems in Fig. 16.5 are designed to save the same lives,
the implementation of duplicate programs would not be desirable even if resources were infinite. However, even nondupli-
cate programs, some considered here and some as yet unana-
lyzed, will never be implemented due to a lack of resources.
The same applies to other social benefits that cannot be
expressed in terms of longevity. As explicitly addressed in
the thoughtful and thorough discussions by R. S. Morison[27]
and T. Gordon's group,[28] the quality of life has many indices
including longevity; and gross inefficiencies result in a
loss to all those indices.

Comments

There are strong incentives to spend greater sums to
improve community safety and health. There are great potential gains. For example, the 1960 age-adjusted white male
mortality in U.S. communities ranges from a low of 850 deaths
per 100,000 population in Tuscaloosa, Alabama, to a high
of 1429 deaths per 100,000 in the Wilkes Barre-Hazelton,
Pennsylvania, community. The extremes in U.S. mortality
range from a low of 499 deaths per 100,000 for white females
in Midland, Texas, to a high of 2037 deaths per 100,000 for
nonwhite males in Erie, Pennsylvania.[3] Past efforts to
improve wellbeing have been fruitful. Even though the
nation's population is growing older, the U.S. mortality
rate in 1975 dipped to 890 deaths for every 100,000, the
lowest in the nation's history, down from 910 in 1974 and
970 in 1968.

We have come to realize, however, that there are substantial costs when one considers spending money on ineffi-
cient programs. These are either the cost: (1) of lives
sacrificed in unfunded but more efficient programs or (2)
of forfeiting alternative benefits from education, housing,
mobility, national security, space exploration, the arts,
or other indices of the quality of life.

References and Notes

1. L. M. Rieser, "The Role of Science in the Orwellian
   Decade," *Science,* 184 (1974), 486.

2. S.H. Preston, N. Keyfitz, and R. Schoen, *Causes of
   Death: Life Tables for National Populations* (Seminar
   Press, New York, 1972).

3. E.A. Duffy and R.E. Carroll, *United States Metropolitan Mortality, 1959-1961* (PHS Publication No. 999-AP-39, U.S. Public Health Service, National Center for Air Pollution Control, Cincinnati, 1967).

4. *Accident Facts: 1974 Edition* (National Safety Council, Chicago, 1974).

5. A. Joan Klebba, "Homicide Trends in the United States, 1900-74," *Public Health Reports*, 90 (1975), 195-204.

6. L.B. Lave and E.P. Seskin, "An Analysis of the Association Between U.S. Mortality and Air Pollution," *Journal of the American Statistical Association*, 68, (1973), 284-290.

7. R.C. Schwing and G.C. McDonald, "Measures of Association of Some Air Pollutants, National Ionizing Radiation and Cigarette Smoking with Mortality Rates," *Science of the Total Environment*, 5 (1976), 139-169.

8. U.S. Office of Science and Technology, *Cumulative Regulatory Effects on the Cost of Automotive Transportation (RECAT)* (Washington, 1972).

9. D. Levine and B.J. Campbell, "Effectiveness of Lap Seatbelts and the Energy Absorbing Steering System in the Reduction of Injuries," mimeographed (Highway Safety Research Center, University of North Carolina, Chapel Hill, 1971).

10. H. Joksch and H. Wuerdeman, "Estimating the Effects of Crash Phase Injury Countermeasures," *Accident Analysis and Prevention*, 4 (1972), 89-108.

11. L. Lave and W. Weber, "A Benefit-Cost Analysis of Auto Safety Features," *Applied Economics*, 2 (1970), 265-275.

12. J.B. Gregory, NHTSA Administrator quoted in *Wards Auto World*, 11 (1975), 10.

13. S. Peltzman, "The Effects of Automobile Safety Regulation," *Journal of Political Economy*, 83 (1975), 677.

14. R. Zeckhauser, "Procedures for Valuing Lives," *Public Policy*, 23 (1975), 419-464.

15. M.C. Weinstein, "Foundations of Cost Effectiveness Analysis for Health and Medical Practices," *New England Journal of Medicine*, 296 (1977), 716-721.

16. R.N. Grosse, "Cost-Benefit Analysis of Health Service," *Annals of the American Academy of Political and Social Science,* 399 (January, 1972), 89-99.

17. W. Stork, "The Cost Effectiveness of International Vehicle Regulations," *Automotive Engineering,* 81 (1973), 35 (Table 4).

18. L.M. Patrick, "Passive and Active Restraint Systems Performance and Benefit/Cost Comparison" (SAE 750389), *Transactions of the Society of Automotive Engineers,* 84, (1975), 10-32.

19. H.E. Klarman, "Cost-Effectiveness Analysis Applied to the Treatment of Chronic Renal Disease," *Medical Care* 6, (1968), 48-54.

20. R.A. Howard, "On Making Life and Death Decisions," in *Societal Risk Assessment: How Safe is Safe Enough?,* ed. R.C. Schwing and W.A. Albers (Plenum, New York 1980), pp. 89-106.

21. H.M. Schmeck, Jr., "Law on Kidney Aid is Termed Unfair," *New York Times,* 29 August 1976.

22. J.P. Acton, *Evaluating Public Programs to Save Lives: The Case of Heart Attacks* (Rand Report No. R-950-RC; Rand Corporation, Santa Monica, 1973).

23. "City with a Heart," *Newsweek* (9 August 1976), p. 77.

24. R.C. Schwing, B.W. Southworth, C.R. von Buseck and C.J. Jackson, "Benefit-Cost Analysis of Automotive Emissions Reductions," *Journal of Environmental Economics and Management,* 7 (1980), 44.

25. *Air Quality Noise and Health: Report of the Interagency Task Force on Motor Vehicle Goals Beyond 1980* (Dept. of Transportation, Office of the Secretary, Washington, 1976).

26. U.S. Dept. of Transportation, *The National Highway Safety Needs Report* (DOT, Washington, 1976).

27. R.S. Morison, "Misgivings About Life-Extending Technologies," *Daedalus,* 107 (Spring, 1978), 211-226.

28. T.J. Gordon, H. Gerjvoy, and M. Anderson, *A Study of Life-Extending Technologies* (NTIS Publication No. PB-279 267; U.S. Department of Commerce, Springfield, VA., 1977), Vol. 1.

Agenda for Research

_____ *Christoph Hohenemser, Jeanne X. Kasperson*

17. Overview:
Two Research Perspectives

As already noted in the introduction, risk analysis is a field with little theory and a wealth of conflicting demands. In the hope that this condition can be improved, we close this volume by presenting two papers that outline substantial agendas for future research. Covello and Menkes write from the perspective of the National Science Foundation Program on Technology Assessment and Risk Analysis (NSF TARA). Kasperson and Morrison present material prepared for the Swedish government through the Beijer Institute, Stockholm.

The two papers are complementary. Taken together, their recommendations in effect propose research leading to: (a) a narrowing of the gap between decision-making and existing risk assessment methodology; (b) an evaluation of the interplay between information flow and risk perception; (c) an examination of neglected risk assessment institutions; and (d) an emphasis on comparative studies. To assist the reader in assimilating the two contributions, we consider each of these directions separately.

Decision-making and risk assessment methodology. Society has made progress in identifying and measuring a wide range of risks. The rub comes in accommodating, regulating, and managing risks. In understanding this difficulty, a key question is how risk assessments are translated into policy and why this process is so often unsuccessful. Both papers advocate work on the role of risk assessment studies in decision-making. Covello and Menkes note that many expensive studies comparing technological risks remain unevaluated, unused, or underutilized in terms of their impact on policy decisions. Kasperson and Morrison cast a critical eye on the so-called "big energy studies" undertaken by various industrial nations and question the extent to which the research findings have made their way into public policy. It may be that at their present state of development, risk assessment studies can influence the ambience in which the decision process operates, but can not, and perhaps should not, translate directly into policy. A case in point may be the Reactor Safety Study, which stimulated the technical side of the nuclear debate as nothing else had, yet eventually was declared by its sponsor, the Nuclear Regulatory Commission, as not relevant to policy.

Information and public perception. More so than most industrialized nations, the United States manages to distribute a maximum of information about technological risks. Not a week passes without the announcement of a new hazard. The source of the information may be one of many government agencies concerned with risk, or one of numerous private organizations with significant risk agendas. In either case, the information is transmitted by the mass media. In the view of some (Covello and Menkes included), well-intentioned efforts to inform the public often backfire and create more fear than anything else. This raises the question of how risk information is transferred and used. Do certain risks receive more play than others, and why? Are underreported risks accepted or acceptable? Should the media, as Okrent (Chapter 13) would have it, bear responsibility for the effects of their risk reports?

To deal with such questions, Covello and Menkes call for analysis of how risk information is disseminated to the general public and to decision-makers. Their aim is to find better methods of communication and to enhance predictive capability. Kasperson and Morrison note that national governments often seek consensus via public-information campaigns and outline a plan for categorizing and assessing such campaigns. They also call for an in-depth study of the media's treatment of risk events, risk estimates, and risk consequences.

In viewing the role of the media, it is important to realize that denouncing the media as scaremongers may be tantamount to blaming the messenger for the bad news. It is also well to remember that newspapers, radio, and television have become the principal ways that people learn about and respond to risks. The morning paper, not the family doctor, tells us about medical research findings on saccharin. Public television, not the electric company, tells us about the risks of the plutonium economy. If we wish to understand society's response to technological risk, the study of information handling and transfer via the mass media could therefore not be much more important. Except for a recent beginning by a committee of the National Academy of Sciences,[1] however, there has been little systematic research on the workings and impacts of the mass media. Once such work gets under way, we may learn more about the source of biases in risk judgment described in Chapter 10 by Slovic and collaborators.

Risk management institutions. To understand why risks are handled and perceived as they are, it is important to look not only to information, but to other institutions. The bulk of work on the latter topic is currently focussed on national regulatory agencies. Although national governments are without question primary risk managers, they have been overemphasized to a degree that shortchanges other subnational and supranational institutions. Both sets of authors recognize and propose to rectify this imbalance and advocate several new directions for research.

Covello and Menkes stress the need to examine American risk management strategies and practices at regional, state, and local levels. Kasperson and Morrison concur but look in addition to work at the micro-level of the local community. One proposed project would analyze labor unions as risk managers, another would examine industry. Both sets of authors also propose studies that transcend the national level in the other direction. Covello and Menkes argue that the experiences of other countries will provide useful international comparative data, and Kasperson and Morrison envision reciprocal benefits arising from cross-national comparisons and international collaboration.

Comparative studies. The call to consider neglected risk management institutions leads naturally to a concern with integrated and comparative risk assessments. Technological hazards show no respect for national boundaries. Oil spills, acid precipitation, carbon dioxide buildup all trespass where they will, often far from their places of origin. National and international responses, or lack of them, carry far-reaching implications. A recent report has indicated that national self-interest still outstrips collaborative handling of environmental risk.[2] Indeed, it is difficult to conjure up a national government that would willingly pay only to protect the health of another country.

International collaboration on risk questions at the very least should facilitate the sharing of information, scientific knowledge, and management experience. The four authors of this part advocate a utilitarian, almost simplistic, learn-by-example approach, involving trust, data sharing, and willingness to accept the findings of others.

Kasperson and Morrison propose the compilation of risk profiles for developing countries and the exchange of information to fill and identify gaps in knowledge about specific risks. A special case involves full disclosure by advanced countries of information on hazardous technology that is transferred to developing nations.

Both pairs of authors in this part value the compilation of case studies, as models or paradigms and as predictive and anticipatory tools for cross-hazard and cross-national comparison. Covello and Menkes suggest that comparable case studies documenting differences in handling of the same risk by various countries may serve to inform American public policy. One of the proposals of Kasperson and Morrison is to compare national responses to energy risks; another would compile case studies on the export, and thereby on the corresponding import, of hazardous products and technologies. Both sets of authors see the need to choose cases that are illustrative of effective as well as ineffective risk management, and both deplore past concentration on failure. In fact, Kasperson and Morrison call for "best-case" studies of industries, with emphasis on competence instead of ineptitude.

The agenda for further research is therefore rich and formidable. It looks for the practical, the useful, and the relevant. It also seeks a unified basic understanding where it can. Above all, it draws its strength from the many urgent and immediate problems that society has with the hazardous side effects of technology.

References and Notes

1. *Disasters and the Mass Media: Proceedings of the Committee on Disasters and the Mass Media Workshop, February, 1979* (National Academy of Sciences, Washington, 1980).

2. *Environmental Risk Assessment,* ed. A.V. Whyte and I. Burton (SCOPE 15; John Wiley, for the Scientific Committee on Problems of the Environment (SCOPE) of the International Council of Scientific Unions, New York, 1980), p. 134.

Vincent T. Covello, Joshua Menkes

18. Issues in Risk Analysis

Introduction

In recent years, public attention has focused on an array of technological hazards that pose a significant threat to human health, safety, and the environment. At the same time, governmental responsibilities for technological hazard assessment and management have grown because of a perceived need to anticipate, prevent, or reduce the risks inherent in modern society. In attempting to meet these responsibilities, legislative, judicial, and regulatory bodies have had to deal with the extraordinarily complex problem of assessing and balancing risks, costs, and benefits.

The need to help society cope with technological hazards has given rise to the development of a new intellectual endeavor: risk analysis. The scope and complexity of risk analyses require a high degree of cooperative effort on the part of specialists from many fields. Analyzing technical, social, and value issues requires the efforts of physicists, biologists, geneticists, physicians, chemists, engineers, political scientists, sociologists, decision analysts, management scientists, economists, psychologists, ethicists, and policy analysts.

The following sections briefly outline risk analysis issues warranting further exploration and discussion. For a more comprehensive review of past work in this field, the interested reader is referred to the growing body of

---

The views expressed in this paper do not necessarily represent the views of the National Science Foundation, but are exclusively those of the authors.

literature on risk and its management.[1] Much of this work draws upon a larger literature on decisionmaking, probability theory, and risk that has developed over the last fifty years.[2]

Policy Issues

Research on risk can aid policy decisionmakers and the public by laying out technical, social, and value issues and by framing issues so that they can be readily comprehended. Some basic dimensions of the problem may be phrased as follows:

(1) How safe is safe enough? Assuming that it is possible to arrive at more or less accurate descriptions of both risks and benefits, how should socially acceptable levels of risk be determined? To what extent should socially acceptable levels of risk vary according to estimates of potential benefits? Are some risks unacceptable no matter what the expected benefit?

(2) How adequate are the data on which we depend for estimates of the risk associated with various technologies, and how can such adequacy be judged? What can be done to improve the data, or at least to develop more precise understandings of the confidence levels associated with the data? What mechanisms can be developed to ensure periodic reassessments of risks based on new knowledge?

(3) What strategies can be developed for coping with situations in which the risk is essentially unknown, and perhaps unknowable?

(4) How are implicit estimates of risk used in the decisionmaking process, both in the public and the private sector? Are there, and should there be, differences in the way public and private sector organizations use risk analyses in decisionmaking?

(5) What are the institutional constraints associated with decisionmaking involving risk and uncertainty? Since most analyses of risk and related decisions are performed in an organizational context, what organizational characteristics inhibit or facilitate effective technological risk management?

(6) What factors influence individual and social perceptions of risk? What is the relation between individual perceptions of risk and social perceptions of risk? What accounts for apparent anomalies in the way

members of society behave when presented with information about risk? What factors account for changes in perceptions of risk? Are there ways to increase the capacity of both the public and decisionmakers for dealing with issues involving risk and uncertainty in a rational manner? Can we improve methods for communicating information about risk?

(7) How are value, equity, distributive, and intergenerational considerations balanced in the decisionmaking process? What means can be used to balance the costs, risks, and benefits to different groups within the population? To what extent should victims of technological hazards be compensated for their losses? Can improved methods be identified for comparing and integrating information about risks so that citizens and policymakers can make well-informed decisions?

Current Knowledge

In the following sections, major research findings related to these issues are briefly described.

Risk Identification, Estimation, and Assessment

Identification. Some hazards are readily identified because their consequences are direct and unambiguous. Others, such as those caused by minute quantities of chemicals, viruses, or radiation, can be detected only by careful research, screening, and monitoring. Particularly difficult is the **a priori** identification of hazards whose consequences appear only after years or decades. For example, the effects of exposure to carcinogens, manifested by increased cancer incidence, may take 15 to 40 years to emerge as a clear trend.

Enumeration and Assessment of Consequences. The basis of a risk analysis is the enumeration and assessment of the consequences of a hazardous event. Unfortunately, no reliable and valid method exists for specifying the consequences of a hazard. As several researchers have shown,[3] even in the most sophisticated analyses important consequences are occasionally omitted.

Probability Assessment. Once the possible consequences of a hazardous event have been identified and enumerated, their likelihood must be assessed. Such assessment is particularly difficult in the case of rare events, since adequate statistical data are often lacking or even uncollectable. Although analysts have developed numerous methods, such as fault tree and event tree analyses, to circumvent this difficulty, many critics question whether such probabilities can be adequately assessed. Holdren,[4] for example, maintains that probability values as small as one chance in 100,000 or one chance in a billion are meaningless and cannot be supported by convincing theoretical arguments, since it is highly likely that important failure modes have been omitted from the analysis. Problems in assessing low probabilities have also been addressed by other authors.[5] What is needed is a careful state-of-the-art review assessing alternative methods and their advantages and disadvantages, as well as a critique of the uses and abuses of probability assessments in previous studies.

Risk Evaluation

The Perception of Risk. Attitudes toward risk are affected by the way individuals perceive risks and benefits and by their ability to comprehend probabilistic statements.[6] Several general conclusions are suggested by recent research on how individuals perceive and process probabilistic information.

First, individuals find it difficult to make decisions about activities involving risk and may not be intellectually or emotionally equipped to respond constructively to the difficulty. Some of the ways individuals systematically violate commonly accepted principles of rational decisionmaking when asked to evaluate probabilities and risks have been documented in a recent study.[7] The need to reduce cognitive strain and anxiety often leads to unrealistic oversimplifications of essentially complex problems and to "solutions" that are more apparent than real.

Second, there is some evidence[8] that the mere possibility of a catastrophic event is not the only consideration in an individual's perception of the severity of a risk. Other important considerations are whether the exposure to risk is voluntary or involuntary and whether the deleterious consequences are immediate or delayed.

Third, individuals do not always perceive risks accurately or act consistently when faced with ostensibly comparable risk situations. They tend to overestimate hazards that are easy to imagine or recall, have particularly dread consequences, are certain to produce death rather than injury, and take several lives instead of one. Conversely, they tend to underestimate the risks of common hazards that kill or injure one person at a time, even if such consequences occur with high overall frequency.[9]

Fourth, problems of misperception are aggravated by the fact that once beliefs are formed an individual will tend to structure and distort the interpretation of new evidence, thereby creating high resistance to information that confirms an opposing view or discredits one's own beliefs. People tend to dismiss evidence contradicting their beliefs as unreliable, erroneous, and unrepresentative.[10]

Finally, particular risk perceptions may be affected by the way information is presented. For example, risks from radiation may seem negligible when described in terms of average reduction in <u>life expectancy</u> for the population within a given radius of a nuclear power plant; however, those same risks may appear substantial when translated into the number of additional <u>cancer</u> deaths per year. Furthermore, reassuring statements by technical experts may do little to alleviate fear. Indeed, informative discussions of a possible disaster, even to demonstrate its improbability, may be counterproductive if they enhance the perception of its reality and facilitate "thinking about the unthinkable."

<u>The Acceptability of Risk</u>. A basic question in risk analysis is whether a product or technology is "acceptably" safe (acceptance denoting anything from embracing the product or technology, to passive acquiescence, to fatalistic submission). To answer the question, it is necessary to know "how safe is safe enough." People are frequently inconsistent in their decisions about what risk levels are acceptable. For example, some communities that live placidly along geologic faults or below great dams may actively protest against the risks of a local nuclear reactor. Such apparent inconsistencies may reflect incomplete knowledge, faulty decisions, and haphazard adjustment, or they may reflect deeper attitudes toward specific hazards. So far two approaches--revealed and expressed preferences--have been used to assess what is acceptable risk.

The revealed preferences method is based on the assumption that historical data reveal patterns of tradeoffs between acceptable risk and benefit. Acceptable risk for a new technology is assumed to be comparable to the risk associated with existing activities having a similar benefit to society. Using this method, Starr[11] discovered that (a) the acceptability of risk from an activity is proportional to the benefits; and (b) the public distinguishes between risks arising from voluntary activities (such as skiing) and risks arising from involuntary activities (such as the consumption of food containing preservatives).

The expressed preferences method is more straightforward. Individuals are asked to state directly what they find acceptable. The method has obvious appeal, for it elicits current preferences and is thus responsive to changing values. It also permits substantial citizen participation in decisionmaking and should therefore be politically acceptable. Nevertheless, it has some drawbacks. People may not really know what they want; their attitudes may be inconsistent with their behavior; their values may change so rapidly that systematic planning becomes impossible; they may not understand how their preferences will translate into policy; they may prefer alternatives that are not realistically obtainable; and they may state different preferences when the same question is phrased in various ways.

Rowe[12] suggests a model that incorporates both the revealed and expressed preference approaches for determining acceptable risks for any given activity. He proposes starting with risk levels reflected in historical data, and then adjusting these levels according to the inequities induced by a particular activity (i.e., risks to individuals other than the ones who reap the benefits) and the controllability of the risk. Greater inequity and less control mean lower permissible risk.

Considering the state of the art in this area, a critical evaluation of the usefulness of the revealed and expressed preference methods is needed. Such a critique should also examine the philosophical underpinnings of each method.

The Value of a Life. A frequently debated issue deals with placing a monetary value on the loss of a human life. Different Federal agencies use vastly different estimates of such a loss. Recently this issue has come to be viewed as placing a value on saving lives or reducing the probability of death. Again, no accepted method exists. Revealed and expressed preference methods that have been advanced for this purpose have been heavily criticized on both empirical and philosophical grounds. Nonetheless, advocates of a decision-analytic approach to the problem argue that a quantitative value is necessary and that a meaningful one can be found. Clarification of this issue, particularly as it applies to the value placed on deaths or injuries in future generations, is needed.

Decisionmaking Methodologies. A variety of methods--such as benefit-cost analysis, benefit-risk analysis, and decision analysis--are available to help decisionmakers evaluate risks, estimate benefits, and deal with uncertainty. All these methods are based on maximizing expected utility and therefore overlap considerably but differ in mode of execution. Several methodological shortcomings can be noted. First, the basic data often do not lend themselves to precise quantification. Examples are the vagaries of nature (where and when a tornado will strike), the potential for human operating error (as in a nuclear power plant), the risk from deliberate abuse of technology (as in terrorism), and the difficulty of dealing with hazards which involve long chains of delayed and currently nonquantifiable (and possibly nonexistent) effects on health. Second, the consequences of alternative control decisions are hard to specify. Third, it is difficult to incorporate values into decision models, particularly where value conflicts exist. Fourth, it is not clear how costs, benefits, and risks to future generations should be treated. Finally, as Stokey and Zeckhauser[13] have noted, benefit-cost analysis is particularly vulnerable to misapplication through carelessness, naivete, or outright deception.

Risk and the Law. Disputes involving risk often end up in litigation, yet many classic legal concepts are difficult to apply in specific cases. Terms, such as "forseeability," "negligence" and "imminent hazard," which once had relatively rigid meaning, are often inadequate for dealing with the uncertainties associated with risk analysis. Risk analysis often must rely on incomplete scientific evidence and on probabilistic reasoning, resulting in notions of proof of causality that are only tentative. As a result, the courts are seeking more appropriate notions of what such terms should mean and how they should be applied to specific cases.[14]

It is indicative of this dilemma that in 1980 a sharply divided Supreme Court invalidated the Occupational Safety and Health Administration (OSHA) standard for worker exposure to benzene, a widely used industrial chemical that is known to cause cancer at high levels of exposure. In a 5-to-4 decision, the Court said that OSHA had not made a proper initial finding on which to base the benzene standard. Nor, it said, had OSHA proved that an older, less stringent benzene standard is inadequate. In handing down the decision, the Supreme Court did not address the more complex issue of whether an agency is required to weigh the benefits of safety standards against the costs of compliance. The United States Court of Appeals had ruled

Table 18.1.   Major Congressional Legislation to Regulate Technological Hazards, 1966-1980.

National Traffic and Motor Vehicle Act (1966)
Oil Pollution and Hazardous Substances Control Act (1966)
Federal Food, Drug, and Cosmetic Act Amendments (1968)
National Environmental Policy Act (1969)
Federal Mine Safety and Health Act (1969)
Occupational Safety and Health Act (1970)
Federal Railroad Safety Act (1970)
Clean Air Act (1970)
Poison Prevention Packaging Act (1970)
Federal Insecticide, Fungicide, and Rodenticide Act (1972)
Marine Protection Research and Sanctuaries Act (1972)
Consumer Product Safety Act (1972)
Ports and Waterways Safety and Health Act (1972)
Water Pollution Control Act (1972)
Noise Control Act (1972)
Safe Drinking Water Act (1974)
Toxic Substances Control Act (1976)
Resource Conservation and Recovery Act (1976)
Solid Waste Disposal Act (1976)
Clean Air Act Amendments (1977)
Clean Water Act (1977)
Hazardous Liquid Pipeline Safety Act (1979)

that such a benefit-cost analysis was required when it struck down the new benzene standard in 1978. OSHA had appealed this decision, arguing that the benefits and costs of a toxic substance often cannot be quantified.

In addition to these problems, disputes over individual rights versus societal rights raise profound ethical and legal issues that are difficult to resolve either conceptually or in practice. Some issues include the right of terminal cancer patients to treatments of their choice (e.g., laetrile) versus society's right to protect itself by discouraging and punishing quackery; the right of motorcyclists not to wear helmets versus society's right to reduce demands on public emergency facilities by protecting individuals against their own folly; the right of the Amish to deny vaccination to their children versus society's right to protect itself against reservoirs of infection.

Current Problems in Federal Regulation

Public concern over technological hazards has led to substantially increased government involvement in risk management. In the last two decades, for example, at least 23 major pieces of legislation to regulate technological hazards have been passed by the U. S. Congress (see Table 18.1). This increase in legislative activity has been paralleled by rapid expansion of staff, scientific capabilities, and monitoring programs in the various Federal regulatory agencies.

The increasing emphasis on government regulation of technological hazards raises a number of important problems: First, laws and regulations are often inconsistent (e.g., allowing carcinogens in the air but not in food). Second, government agencies often use different approaches to risk management (e.g., the Food and Drug Administration and the Environmental Protection Agency). Third, as Kates[15] points out, it is extremely difficult to design appropriate standards and regulations to govern the multitude of forms in which a hazard appears. The magnitude of the task is exemplified by the Environmental Protection Agency's attempt to formulate generic standards for the 40,000 local public water supplies in the United States. Fourth, if a manufacturer does not cooperate or provides inadequate safety tests, the regulator may become extensively involved in the collection and evaluation of data to the point that the magnitude of the task overwhelms the resources of the agency. Fifth, the economic costs of regulation sometimes exceed the economic benefits. For example, a poorly conceived regulation may blunt industry's research and development, impede technological growth, disrupt the configuration of a particular industry (e.g., drive out small producers), and adversely affect the market competitiveness of a product. In such circumstances, insurance may offer an attractive alternative to regulation. Sixth, regulating a particular product or process may inequitably allocate the costs, risks, and benefits among various subpopulations. Even if the aggregate benefit-cost ratio is favorable, it is not clear how distributive impacts should be weighed and incorporated into the decision process. Finally, many observers of governmental regulation argue that increased reliance on regulatory solutions is ultimately detrimental to society's management of hazards, and that greater reliance should be placed on providing more and better information to the public, on personal good judgment, and on after-the-fact claims for damages. Is this a preferable mode of hazard management? If so, for what type of hazard? What effect should high degrees of uncertainty or varying technological complexity have for the preferred mode of hazard management?

Conclusion

As part of the effort to resolve these problems, researchers can contribute information that will be of use in the formulation of risk management policies. To this end, detailed analyses of particular risks will continue to be needed. Of equal importance is risk analysis research undertaken to improve (a) communication of information about risk to decisionmakers and the lay public; and (b) understanding of how to balance distributive, equity, and other normative considerations in the policy decisionmaking process. In order to achieve these objectives, the foregoing review suggests that research on at least five topics is needed.

Risk Assessments and Their Impact on Risk Management Policies. During the last decade, repeated attempts have been made to assess and compare the risks associated with various technologies. These efforts have been substantial undertakings, involving large commitments of scientific and financial resources from the public and private sectors. The impacts of these assessments on risk management policies should be determined through detailed institutional and organizational analyses. The products of such analyses would help to pinpoint the difficulties involved in translating assessments into social policies.

Risk Assessment and Management at the Regional, State,

and Local Level. Many risks are managed at the regional, state, and local level. Yet relatively little is known about what risk management approaches are used, what risk assessment capabilities exist, and how regional, state, and local authorities vary in their implementation of Federal risk regulations. Studies are needed that will fill this informational void. The objective of such studies is to identify means by which regional, state, and local risk assessment and management can be improved.

International Comparisons. With proper allowance for cultural and institutional differences, U.S. risk management policy could benefit from analyses of the experiences of other nations. Unfortunately, too little effort has been made to compare risk management policies and practices of the U.S. with those of other nations, or to compare successes and failures in these efforts. Studies are therefore needed to examine (a) the extent to which nations respond selectively to various technological risks; (b) the degree of consensus or disagreement in their evaluations of technological risks; (c) the interaction of risk management policies and social, economic, political, and institutional factors; and (d) the effectiveness of alternative managerial strategies in reducing or mitigating risks. The products of such analyses should be specifically designed to answer the question: What improvements in U.S. risk management policies and practices are suggested by international comparisons?

Public Perception of Risk. Behavior toward risk is affected by perceptions of risk and by the ability to understand the consequences of exposure to particular hazards. Although issues in risk perception are an important element in the risk management decisionmaking process, comparatively little is known about the cognitive processes that determine the perception of risk and the ordering of preferences. Studies are therefore needed that would examine (a) the psychological, social, and institutional factors affecting risk perception and choice; (b) the conditions or events under which risk perceptions and preferences remain stable or change; (c) how information about risk is communicated to and understood by decisionmakers and the lay public; and (d) the process by which individual perceptions of risk are aggregated and translated into social perceptions of risk and decisionmaking. The objectives of these studies are both to identify better means for communicating information about risk and to develop improved methods for predicting public response and involvement in risk management decisionmaking.

Decisionmaking Methods. Methods currently used to
evaluate technological hazards have serious limitations.
The basic data are often inadequate, the consequences of
decisions are hard to specify, it is difficult to
incorporate equity and other normative considerations into
decision models, there is no agreement on how to deal with
uncertainty or value of life issues, and it is not clear how
costs, benefits, and risks to future generations should be
treated. Studies are therefore needed that would examine
(a) the extent to which decisionmaking methods have been
used in risk management; (b) successes and failures in the
applications of these methods; and (c) major unresolved
methodological needs and prospects for resolving them in the
near future. The objectives of these studies are to clarify
the policy usefulness of currently available methods and to
assess the need for new or improved methods.

A major research effort focused on these five topics
should lead to greater understanding and more systematic
thinking about risk. This is not to say, however, that risk
management problems will be solved. Although some risks can
be deliberately managed, others will remain intractable.
For many hazards, an understanding of causality will elude
us for the foreseeable future, and we must therefore decide,
in the presence of continuing, substantial uncertainties,
which risks to live with and which to reduce.

References and Notes

1. W.D. Rowe, *An Anatomy of Risk* (John Wiley, New York,
    1977); D. Okrent, ed, *Risk-Benefit Methodology and Ap-
    plication: Some Papers Presented at the Engineering
    Foundation Workshop, Asilomar* (UCLA-ENG 7598; University
    of California, Department of Energy and Kinetics, Los
    Angeles, 1975); National Academy of Engineering, Commit-
    tee on Public Engineering Policy, *Perspectives on Ben-
    efit-Risk Decision Making* (National Academy of Sciences,
    Washington, 1972); R.W. Kates, ed., *Managing Technolog-
    ical Hazard* (University of Colorado, Boulder, CO., 1977);
    A. Tversky and D. Kahnemann, "Judgment under Uncertain-
    ty: Heuristics and Biases," *Science,* 185 (1974), 1124-
    1131; P. Slovic, B. Fischhoff, and S. Lichtenstein,
    "Rating the Risks," *Environment,* 21 (April, 1979), 14-
    20, 36-39 (also Chapter 10, this volume); C. Starr,
    "Social Benefit versus Technological Risk," *Science,*
    165 (1969), 1232-1238; H.J. Otway and J.J. Cohen, *Re-
    vealed Preferences: Comments on the Starr Benefit-Risk
    Relationship* (RM-75-7; International Institute for Ap-
    plied Systems Analysis, Laxenburg, Austria, February
    1975); R.C. Schwing and W.A. Albers, eds., *Societal*

Risk Assessment: How Safe is Safe Enough? (Plenum Press, New York, 1980); A. Wildavsky, "No Risk is the Highest Risk of All," *American Scientist,* 67 (January/ February, 1979), 32-37; W.W. Lowrance, *Of Acceptable Risk: Science and the Determination of Safety* (William Kaufman, Los Altos, California, 1976); National Research Council, Committee on Principles of Decision-Making for Regulating Chemicals in the Environment, *Decision-Making for Regulating Chemicals in the Environment* (National Academy of Sciences, Washington, 1975); I. Burton, R.W. Kates, and G.F. White, *The Environment as Hazard* (Oxford University Press, New York, 1978); B. Fischhoff, "Cost-Benefit Analysis and the Art of Motorcycle Maintenance," *Policy Sciences,* 8 (1977), 177-202; C. Starr and C. Whipple, "Risks of Risk Decisions," *Science,* 208 (1980), 1114-1119 (also, Chapter 14, this volume; R. Wilson, "Analyzing the Risks of Daily Life," *Technology Review,* 14 (February, 1979), 40-46.

2. J.M. Keynes, *A Treatise on Probability* (Macmillan, London, 1921); J. von Neumann and O. Morgenstern, *The Theory of Games and Economic Behavior* (Princeton University Press, Princeton, NJ, 1944); F. Knight, *Risk, Uncertainty, and Profit,* Reprints of Economic Classics (Augustus M. Kelley, New York, 1964); L.J. Savage, *The Foundations of Statistics* (John Wiley, New York, 1954); M. Friedman and L.J. Savage, "The Utility Analysis of Choices Involving Risk," *Journal of Political Economy,* 56 (1948), 279; H.A. Simon, "A Behavioral Model of Rational Choice," in H.A. Simon, *Models of Man* (John Wiley, New York, 1957), p. 241; H.A. Simon, "Theories of Decision-Making in Economics and Behavioral Science," *American Economic Review,* 49 (1959), 279; K.J. Arrow, "Utility and Expectation in Economic Behavior," in S. Koch, ed., *Psychology: A Study of a Science,* Volume 6: *Investigations of Man as Socius: Their Place in Psychology and the Social Sciences* (McGraw-Hill, New York, 1963), p. 724; K.J. Arrow, *Aspects of the Theory of Risk-bearing* (Academic Press, Helsinki, Finland, 1965); R.D. Luce and D.H. Krantz, "Conditional Expected Utility Theory," *Econometrica, 39* (1971), 253.

3. R.W. Kates, ed. *Managing Technological Hazard* (University of Colorado, Boulder, CO., 1977); B. Fischhoff, "Cost-Benefit Analysis and the Art of Motorcycle Maintenance," *Policy Sciences,* 8 (1978), 177-202.

4. J.P. Holdren, "The Nuclear Controversy and the Limitations of Decision Making by Experts," *Bulletin of the Atomic Scientists,* 32 (March 1976), 20-22.

5. W.G. Fairley, "Criteria for Evaluating the 'Small' Probability of a Catastrophe Accident from the Main Transportation of Liquefied Natural Gas," in *Risk-benefit Methodology and Application: Some Papers presented at the Engineering Foundation Workshop, Asilomar,* ed. D. Okrent (UCLA-ENG 7598; University of California, Department of Energy and Kinetics, Los Angeles, 1975); B. Fischhoff, "Cost-Benefit Analysis and the Art of Motorcycle Maintenance," *Policy Sciences,* 8 (1978), 177-202; A.E. Green and A.J. Bourne, *Reliability Technology* (Wiley-Interscience, New York, 1972); P. Slovic, B. Fischhoff, and S. Lichtenstein, "Behavioral Decision Theory," *Annual Review of Psychology,* 28 (1977), 1-39; A. Tversky and D. Kahnemann, "Judgment under Uncertainty: Heuristics and Biases," *Science,* 185 (1974), 1124-1131; C. Starr, R. Rudman, and C. Whipple, "Philosophical Basis for Risk Analysis," *Annual Review of Energy,* 1 (1976), 629-662.

6. The material in this section is, for the most part, taken from a report on three workshops on risk assessment and societal response to environmental hazards of human origin, supported by the National Science Foundation (see R.W. Kates, ed., *Managing Technological Hazard,* University of Colorado, Boulder, CO., 1977).

7. A. Tversky and D. Kahnemann, "Judgment under Uncertainty: Heuristics and Biases," *Science,* 185 (1974), 1124-1131.

8. B. Fischhoff, P. Slovic, S. Lichtenstein, B. Comba, and S. Read, "How Safe is Safe Enough? A Psychometric Study of Attitudes Toward Technological Risks and Benefits," *Policy Sciences,* 8 (1978), 127-152.

9. H. Kunreuther *et al., Disaster Insurance Protection: Public Policy Lessons* (Wiley-Interscience, New York, 1978); P. Slovic, B. Fischhoff, S. Lichtenstein, "Cognitive Processes and Societal Risk Taking," in *Cognition and Social Behavior,* ed., J.S. Carroll and J.W. Payne (Lawrence Erlbaum Associates, Potomac, MD., 1976).

10. D. Nelkin, "The Role of Experts in a Nuclear Siting Controversy," *Bulletin of the Atomic Scientists,* 30 (1974), 29-36; L. Ross, "The Intuitive Psychologist and His Shortcomings: Distortions in the Attribution Process," unpublished manuscript (Stanford University, Palo Alto, n.d.).

11. C. Starr, "Social Benefit versus Technological Risk," *Science,* 165 (1969), 1232-1238.

12. W.D. Rowe, *An Anatomy of Risk* (John Wiley, New York, 1977).

13. E. Stokey and R. Zeckhauser, *A Primer for Policy Analysis* (W.W. Norton, New York, 1978).

14. D.L. Bazelon, "Risk and Responsibility" *Science,* 205 (1979), 277-280; D.L. Bazelon, "Coping with Technology Through the Legal Process," *Cornell Law Quarterly,* 62 (1977), 817; M.L. Cohen, J. Stepan, N. Ronen, *Law and Science: A Selected Bibliography* (Harvard University, Cambridge, MA., 1978); "Hazardous Substances in the Environment: Law and Policy," Special Issue of *Ecology Law Quarterly,* 7 (1978).

15. R.W. Kates, ed., *Managing Technological Hazard* (University of Colorado, Boulder, CO., 1977).

Roger E. Kasperson, Murdo Morrison

19. A Proposal for International Risk Management Research

Although there is a rapidly growing scientific effort in risk research, it is fragmentary and uncoordinated. Methodology of risk assessment, despite the lack of quantitative achievement noted by Okrent (Chapter 13) has far outstripped the ability to incorporate its results in decision-making. In fact, there are important gaps in our knowledge of how such judgments and decisions are actually made. There has been a general lack of cross-national research despite the clear need to learn from the experience of others, to define strategies for societal response to risks, and to identify effective institutional arrangements. Governmental handling of hazards, often ad hoc and case-by-case, highlights the need for risk taxonomies (cf. Hohenemser et al., Chapter 9; Gori, Chapter 11) that incorporate attributes of hazards and sociopsychological response. Nearly all research has focussed on federal governmental regulators; precious little is known of the roles of industry, labor unions, the mass media, and local communities. Scientific effort has also concentrated on advanced industrial societies to the neglect of risk issues in developing societies. At the logistical level, although a number of major research groups exist in Europe and the United States, few have ensured long-term research support and linkages are sporadic.

An International Research Program

The options for risk management research are quite numerous for a field in such an early state of development. Most advanced industrial societies are struggling to formu-

---

Adapted from a working paper of the Beijer Institute, Stockholm, Sweden.

late more generic risk management approaches, whereas Third World countries confront balancing between developmental and environmental protection objectives. Risks of various types are very much at the center of discussion and debate. The past decade has witnessed substantial efforts to identify and assess the risks posed by alternative energy systems, and there is widespread current concern over chemical risks. Although active controversy may be expected to continue for some time, it is apparent that the ability to define risks has outstripped the ability to act upon that knowledge. While further work, of course, is needed in identifying, quantifying, and comparing these risks, the more substantial difficulties appear to lie in translating these assessments into effective social actions.

The 10-project research program suggested here addresses three major needs. First, the increasing role of the public in risk policy-making requires greater attention to the problems involved in communicating risks to lay people and anticipating issues of public concern. There is similarly a need for greater understanding of the public response to various risk situations to guide the actions of those who make daily decisions. Given the influential role of the mass media, in both everyday and disaster situations, in shaping lay opinions and setting political agendas, analysis is needed of their adequacy in providing balanced and accurate information. Second, risk research to date has centered nearly exclusively upon the activity of national governmental institutions. Yet a variety of other institutions (e.g., industry, labor unions, local communities) play essential roles in the social management of risk more broadly defined. Finally, greater understanding of risk in comparative and international terms is a prerequisite for improved managerial performance. The proposed program addresses five aspects: the impacts of large-scale risk assessments, the effectiveness of alternative national responses and institutions for risk management, national profiles of risk generation and/or experience, the export of technological hazards, and, finally, opportunities for internationally coordinated risk management.

A standard format is used in presenting this research program. Following an introduction of each risk research problem is a detailed project description that notes the research objectives and specifies in brief the approach or procedures to be followed. The selection of projects involved two major considerations: the contribution to the overall scientific understanding of risk management, and the contribution to improved risk management practice over the near term. Table 19.1 provides a summary of the program.

Risk Communication and
Public Responses

Project #1: Risk Presentation in the Mass Media

The nature of technological risks, often associated with complex technology, multiple emission pathways, and poorly understood chronic effects, hampers the accurate communication of information to the public. Most difficult risk cases involve substantial uncertainty, probabilistic thinking, disagreements among experts, complex fault mechanisms, and a variety of often dimly perceived ecological and health consequences. Low-probability/high-consequence events appear particularly difficult for lay persons to assess. These issues, of course, challenge efforts at public education, information dissemination, and the building of social consensus. The presentation of risks in the mass media substantially influences the response of politicians and laypersons alike, for it constitutes a significant part of our experience, albeit indirect, with risk events.

There have been relatively few useful systematic studies evaluating risk descriptions in the newspapers, much less other media. The Swedish Risk Generation and Risk Assessment Project, however, included a suggestive study of risk coverage in Swedish newspapers.[1] Mass press and media analyses have been conducted by CEPN (Centre d'Etude de la Protection dans la domaine Nucleaire) for France.[2] A recent study by Decision Research in the U.S. found that all forms of disease were greatly underreported whereas violent and catastrophic events (tornadoes, fires, homicides) were over-reported.[3] Although biased newspaper coverage may serve a useful function in alerting society to the need for corrective action for certain risks, it may also create a need for informational programs to counter the distorted public perceptions that result.

A special problem is presented by disasters when the usual sources and channels of information break down or reliable data are not readily available to mass media representatives. The record of performance appears mixed. The mass media failed to produce adequate warning during the 1977 Andhra Pradesh cyclone, but reporters from Le Monde and the New York Times were instrumental in alerting world attention to the Sahel drought.[4] A comparative analysis of coverage by four Canadian newspapers of six risk events found considerable confusion and inaccuracy.[5] The U.S.

Table 19.1 Summary of the proposed research program

Project	Objectives	Approach
1. Risk Presentation in the Mass Media	Identify trends in risk coverage; assess accuracy; analyze internal journalistic processes; note failings; present recommendations	Factual analysis of current trends in risk presentation
2. Public Information and Consensus Building	Develop information base; categorize and assess campaigns; relate findings to other research; present recommendations	Longitudinal study of campaign; comparative analysis of effectiveness of past campaigns
3. Public Attitudes Toward Intolerable Risk	Define public perceptions, relate to risk; suggest means for incorporating into public policy; provide recommendations	International workshop to develop future research; longitudinal studies
4. Industrial Risk Management: An Exploratory Study	Evaluate technical competence; examine profit/safety conflicts; suggest means for transferring success	Overview of industrial risk assessments; "best-case" analysis
5. Labor Unions and Occupational Risk	Analyze role of unions in risk reduction; assess monitoring and technical expertise; review past efforts and suggest recommendations	Overview current literature; document contrasts between unions; interviews

Table 19.1, continued

	Project	Objectives	Approach
6.	The Local Community as Risk Manager	Determine what risks handled locally; identify generic problems, suggest research strategy	Field study of local communities in various countries
7.	Assessing the "Big" Energy Assessments	Assess risk clarification, narrowing of debate, impact on public policy; define use by key groups	Review of reports, interview opponents and proponents
8.	Comparing National Responses to Energy Risks	Determine selective responses, effectiveness of institutions and strategies; suggest lessons	Comparative study of several countries; overview major differences; analyze effort
9.	Risk Profiles of Developing Countries	Define major risk dimensions; develop national risk profiles	Categorize risks; plot trends over time (where possible); collect and assess available data
10.	The Export of Risks	Estimate problem magnitude; identify reasons for vulnerability; assess adequacy of response	Case studies

National Academy of Sciences in its assessment of the performance of mass media in disasters found a record of only
minimal completed research and a need for assessing the
accuracy of media reporting during various disaster stages,
the extent of disaster-related information and education programs, and effects on perceptions and behavior.

The project proposed here has four principal objectives:

- to identify trends in risk coverage in the various
  media

- to examine the internal processes of journalism
  which give rise to the portrayal of risks, with
  particular attention to the pressures which limit
  or conflict with more balanced and accurate risk
  presentation

- to assess the scientific accuracy of risk information presented and to note recurring types of
  distortions

- to note specific failings which require attention and to suggest means for improving mass
  media performance

The proposed approach is twofold: factual analysis of
current trends and patterns of risk presentation, and an
assessment of the internal processes of journalism that determine risk portrayal. No effort is proposed to define
public response to the information presented; this might
well be the subject for a follow-up study. A time period
(perhaps a decade) should be selected sufficient to define
the scope of the study. Television, radio, and newspapers
should be included. In regard to newspapers, an effort
should be made to distinguish among dailies, weeklies, and
specialized (such as labor union newspapers) papers. The
scale of information will require statistical sampling procedures.

The analysis of overall trends and patterns would
entail a number of specific tasks:

- selection of means to characterize presentation of
  risk information. This might well include for different risks such indicators as number of deaths
  reported, number of occurrences of risk events,
  number of articles, inches of column space, or
  minutes of air time.

● plotting trends over time in coverage of different
  types of risks and evaluating these trends by type
  of source, type of hazard, and type of technology.
  Differential trends by type of mass media should
  be noted.

● analysis of selected risk events to compare the scientific accuracy of information presented. Again,
  contrasts among the mass media should be identified.

● supplementary studies to determine the technical
  competence of various mass media reporters and analysts and/or their use of other available technical expertise.

Beyond this, detailed interviewing of members of the
mass media would be required. In assessing the internal processes of journalism, a number of questions should be addressed:

● what difficulties do mass media representatives
  encounter in assembling accurate technical information on risks?

● what are the constraints (e.g., time pressures,
  limited technical staff, the need for quick turn-
  around) that limit thorough, accurate, and balanced risk portrayal?

● what internal rules exist for determining what
  is "newsworthy," for providing interpretation
  and analysis, and for selecting among risk can-
  didates for coverage?

Two rather different types of research products would
result. The first would be a series of scholarly reports
on various aspects of mass media presentation of energy
risks, accompanied by a summary statement noting major findings and recommendations for improving mass media performance. The second would be a catalogue of actual mass
media coverage of risks, chosen to illustrate: (1) trends
in energy risk coverage, (2) characteristic types of distortions or failings, (3) characteristic differences in risk
portrayal by types of mass media, and (4) exemplary models
of effective risk presentation.

Project #2: Public Information and Consensus Building

Confronted by deep divisions in the scientific community
and the public over controversial risks (particularly

nuclear power and pesticides), national governments have
instituted a variety of campaigns for informing the public
and for building a consensus for major policy choices. Several European countries, including Sweden, Austria, the
Netherlands and the United Kingdom, have conducted public-
information campaigns on nuclear power.[7] In addition, the
Windscale and Gorleben inquiries[8] have focussed attention
on the problems of nuclear fuel reprocessing and waste disposal. Meanwhile, as Librizzi and Ember point out (Chapters 6 and 7), chemical waste disposal sits on the horizon
as a new and formidable problem. On the one hand, it is
apparent that improved information and public access to
decisions are needed; on the other, effective means for
accomplishing these ends remain elusive. There is substantial preliminary evidence to warn that efforts to provide public information may only serve to raise fears, provide organized opposition with targets of attack, and in
the end make the public more risk averse than before.[9] This
may be the case particularly where low probability/catas-
trophic events are involved.

In appraising the Swedish citizen group discussions of
1974, Dorothy Nelkin suggests that the campaign may have left
the public as uncertain and confused as before,[10] and the
divided results of the 1980 referendum indicate little
social consensus. In the Netherlands, the government hoped
that increased public information would reduce conflicts,
but the results did quite the contrary.[11] The various state
referenda on nuclear power in the United States provide few
reasons to believe that, even with heightened public interest, the marketplace of ideas is an effective vehicle for
increased risk understanding. Even in France a comprehensive public relations program aided by "the resolve of the
French Government without ambiguity" and the support of the
Communist Party have not stifled public concern[12] although
the anti-nuclear movement has lost impetus.[13] The Berger
Inquiry in Canada was perhaps most successful in exposing
a range of far-reaching issues associated with development
of the Alaskan pipeline.[14] The extent of polarization is
quite important, obviously, as evidenced in the Gorleben
inquiry where a panel of "expert" nuclear critics confronted
a group of "expert" pro-nuclear specialists in debate. It
is, as Alvin Weinberg has noted, much more difficult to
"unscare" than to "scare" people.[15] Yet the gap between
public and expert assessment of certain risks is so great
as to require efforts which help the public to put such
risk into meaningful perspective and comparative context.
The fact is, however, that we are quite ignorant as to how
best to accomplish this.

The objectives of the proposed project are:

- to develop a comprehensive information base on various national campaigns, specifically to include assessments of their relative effectiveness for different objectives

- to categorize these campaigns according to their strategies of risk communication and public education

- to assess the actual accomplishments achieved by alternative strategies of risk communication and consensus building, taking account of cultural and institutional differences

- to relate findings to emerging scholarly research on the psychology of risk perception

- to present specific recommendations to be observed by the architects of such informational and educational campaigns

A two-part project is suggested: (a) an intensive longitudinal study of a public information campaign geared to an ongoing risk controversy in a particular country, and (b) a comparative analysis of the design and effectiveness of various past campaigns conducted in a number of countries.

The longitudinal study of a public campaign should examine how the officials who design the campaign conceive of the problem and structure their efforts--their perception of public information and attitudes, objectives recognized, informational strategies chosen, the handling of factual vs. value issues, and criteria recognized for success. Besides monitoring public opinion polls and the eventual referendum results, a panel of respondents (perhaps 25-35 families) might well be selected for continuous, detailed study over the course of the campaign. In addition to reinterviews, it might be useful to have each family keep a diary of information sources, family discussions, and issues of concern. This monitoring should, after establishing baseline knowledge, attitudes, and concerns, chart the impact of the campaign as it unfolds on these families.

The comparative study of campaigns in different countries should provide the context from which to evaluate the experience of any single country. The intent is to specify what can be learned from examining the design and attributes

of information campaigns and relating them to poll results,
opinion surveys, the behavior of organized opposition, and
public response. The outcomes of the campaign would be compared with the initial aims and put into sociopolitical context. To this end, the various national studies of the campaigns should be systematically collected and compared. Conclusions should be drawn as to what elements contribute to
accomplishment of the goals set, their success in achieving
greater public understanding of risk, and the development
of societal consensus. Specific recommendations for improving risk information programs should be offered.

Each research component could result in a major report.
The longitudinal study of a campaign should specifically
relate campaign objectives and strategy to outcomes and
specify lessons to be learned. The comparative study
should provide a summary statement of the implications over
the short and long terms of cross-national efforts for the
tractability of the basic problems.

Project #3: Public Attitudes Toward Intolerable Risk

One of the most perplexing problems confronting risk
management is determining that level at which a particular
risk becomes "intolerable" or, conversely, "tolerable."
It is apparent that the publics in various countries are
more demanding for some types of risks than others. For
example, although many countries have adopted stringent
safety standards for nuclear power and pesticides, the same
countries tolerate high fatality rates from automobiles.
A better understanding of public attitudes to risks, particularly preferences for risk reduction, is essential if
public policies or regulations for technology are to win
broad public support and if decisions are not to be im-
mobilized.

The proposed project would build upon the research
results accomplished in a number of countries, including
particularly the work of Decision Research in the United
States,[16] the International Institute for Applied Systems
Analysis in Vienna,[17] Pieter Stallen in the Netherlands,[18]
Colin Green in England,[19] and Lennart Sjöberg and Ola Svenson in Sweden.[20] These studies suggest that although various publics cannot order risks as accurately as experts,
they are not irrational either. Rather, there appear to be
regular differences between public and expert assessment of
risk, and the gap is particularly large for some risks.
Interestingly, the public apparently consistently overestim-
ates certain well-publicized risks (e.g., botulism, pesti-

cides, nuclear power) while underestimating more serious chronic risks (e.g., smoking, alcoholism). The pattern of public response appears to be related to: (1) objective properties of the risks involved (especially their catastrophic potential, voluntary vs. involuntary nature, newness of the risk), (2) the magnitude of associated benefits, and the correspondence between those benefiting and those at risk, (3) the relative attention devoted to the risk by the mass media, (4) emotional and perhaps deep-seated fears, and (5) other related social questions that impinge upon risk considerations.

The objectives of this project are:

- to bring together representatives of the major research groups in an international workshop to determine major areas of convergence and divergence in research findings and to identify major priorities in needed research as well as possible divisions of labor

- to design a major longitudinal study of perceptions of and attitudes to risk, focussed on formulating underlying theory and assessing the role of causal agents of change

- to suggest means for incorporating public preferences in public policy and regulatory processes and for improving public understanding of risks

At the outset an international workshop should assess what is already known and chart a systematic development of future research to settle existing discrepancies and to advance current theory. Presently, there are promising convergences in findings, but progress is hampered by differing methodologies, varying conceptual assumptions, and infrequent contacts.

Research to date, save Stallen's recent risk survey in the Netherlands[18] and Sjöberg's intensive study of attitudes to nuclear power in Sweden,[20] has primarily involved psychometric experiments. The time is ripe for planning an ambitious longitudinal study of public perceptions of risk aimed at clarifying differences emerging from social and occupational structure as well as the impact of risk events and mass media coverage. Particularly valuable would be an indication of the stability or volatility of such perceptions. Particular attention will need to be given to benefit perception, an area even less researched

and understood than risk perception. Research design questions will be formidable and will require the exploitation
of experimental research findings and the skills of an experienced survey research organization.

Finally, improved understanding is required of how such
information may be more effectively used in the managerial
process. The project should provide a better understanding
of current practices in various countries as well as the
information needs of risk managers. Specific recommendations should be provided for improving public policy.

Non-Governmental Risk Management

Project #4: Industrial Risk Management: An Exploratory Study

Most risk management occurs in the private rather than
the public sector. Industry makes daily decisions in the
development and design of products that affect the exposure
of workers and publics to risks. It is at this early stage,
of course, that the potential for risk control is greatest,
because substantial investments and product dissemination
have not yet occurred. Yet industrial risk management is
presently terra incognita.

Extensive debate wages over ways to mesh the public and
private sectors in risk management. The reliance by some
societies (e.g., the U.K., France) on a cooperative approach
between civil servants and industry has evoked concern over
the underregulation of risk occurring in decisions made
behind closed doors. Others (e.g., the U.S.) employ adversarial processes focussed on formal standard setting, producing fear of over- or inappropriate regulation and needless
conflict. Everywhere questions abound as to how to exploit
best the competence of industry in more effective risk
management.

The objectives of the proposed project are:

● to evaluate the technical competence of selected
   firms within a given industrial sector and suggest needed improvements

● to examine how conflicts between profit and
   safety are resolved in various stages of product
   development

● to identify means for exploiting more effectively
   the technical competence and experience of industry in the societal response to new product risk

● to suggest means for transferring success in one
  sector of industrial risk management to others

To begin the charting of this terra incognita, an
exploratory study is proposed which will define the major
characteristics of the industrial risk assessment process.
This should involve overviewing of the risk assessments performed by industry at the various stages of new product
development. The study should determine the ways in which
risk issues enter into management decisions, whether risk
reduction opportunities are comprehensively and systematically considered, how conflicts between profit and safety
are resolved, whether the firm has a formal safety (such
as product stewardship) policy and the effectiveness with
which it is implemented, and finally, whether safety is
marketed as a product asset. Attention should be given to
characteristics of industrial management that makes for
increased safety.

The accessible literature on industrial risk management
is largely a litany of failure and wrongdoing;[21] successes
command less attention than failures. The project proposed
here is a best-case analysis of industrial risk management.
An international panel should be formed to nominate several
firms particularly noted for their "social responsibility"
in progressive risk management within a given industrial
sector (e.g., automobiles, plastics, contraceptives), as
reflected in product stewardship and in protection of worker
health. It may be particularly useful to incorporate a
cross-national selection of firms--as, for example, Saab
or Volvo, Toyota, General Motors, and Mercedes Benz, in a
study of automobile safety. Institutional differences may
provide added insight into possible improvements in safety
management. The project should report on both generic and
case-study findings, as well as indicating how success may
be transferred to other industrial sectors.

Project #5: Labor Unions and Occupational Risk

There is increasing concern in most industrial societies over the risks present in the work environment. The
predicted toll of asbestos-related diseases has directed
attention to occupational safety as a neglected risk domain.
Characteristically, risks first become apparent in the workplace where they often appear in more concentrated form.
Generally, however, societies tolerate higher exposure
levels in the workplace than in public environments on the
grounds that such risks are either voluntary in nature or
that workers receive compensations for bearing larger risks.
But there is substantial variation in such standards across

countries; the Soviet Union, for example, has more demanding
workplace standards than those in western capitalist countries.[22]

Labor unions have widely varying roles in identifying
risks before worker exposure has become widespread and in
representing worker interests in safer environments. In
Sweden, where the work force is 85 percent unionized, labor
unions elect safety ombudsmen (skyddsombuds) who are present in factories and are empowered to suspend production
temporarily if workers are endangered. The unions in the
United States also engage in extensive safety training and
education. By contrast, labor unions only recently pressed
occupational safety as a major collective agreement consideration, and the worker's right to safety is a controversial
issue. Since labor unions' concern with employment may
sometimes conflict with risk-reduction goals, labor unions
(not unlike industry) must decide between competing goals.

The proposed study would analyze the present role of
labor unions as vehicles for creating safer work environments, comparing accomplishments to date with what is possible. Attention should be given to the activities of unions
in pressing both for risk reduction in specific workplaces
and for broader national legislation on occupational health.
Specific tasks might include:

- an assessment of systematic monitoring or information collection of worker health effects by labor
  unions to provide early identification of risks

- an inventory of cases of risks first identified by
  labor unions, and ways of systematizing early
  warnings of workplace risks

- an analysis of past and present efforts of labor
  unions to reduce occupational risks, both at the
  industry and the society level, with particular
  attention to the effectiveness of safety ombudsmen
  and other innovations

- an examination of the technical capability of unions in risk evaluation, specifically scientific,
  medical, and epidemiological resources

- employing several cross-national case studies,
  an inquiry into how labor unions resolve the
  conflict between the need for employment and the
  need for worker safety

● an evaluation of the extent to which reduction of
occupational risk has become an element in collec-
tive bargaining between unions and management in
different countries and the implications of worker
safety's becoming a bargaining issue

● the formulation of specific recommendations to
increase the effectiveness of labor unions as risk
managers

The project should first provide an overview of the
currently available literature on union activity in risk
protection. Major national contrasts among unions in the
risk area should be carefully documented. Several cross-
national studies of union response to selected occupational
risks would augment the more general analysis of labor union
activity. These cases should be developed in collaborative
research, with a research team in each of the participating
countries. Where available, worker grievances and collec-
tive bargaining agreements would provide useful data for
intensive analysis. Interviews related to the case studies
should be conducted among labor union officials, industrialists, and workers.

Project #6: The Local Community as Risk Manager

Most attention in risk management research has centered upon national institutions and issues. Yet risks are
experienced by persons in local communities and regions. In
such localities individual citizens can play a more active
role in public participation and risk decision-making.
Clearly many risks are already now handled, or may be more
amenable to handling, at the regional or local community
level, yet relatively little is known as to exactly what
these risks are, what local capability exists for their
control, and how successful existing management is.

In Sweden, local communes (the basic local government
jurisdiction) are currently involved in several ways as energy risk managers:

● each commune is presently responsible for the pre-
paration of energy plans, in which risk control
is one consideration

● communes play a role in rate-setting for energy use
and sometimes are directly responsible for the distribution system

- in their choice of heating plants, communes must consider relative risk to local inhabitants within the context of national emission standards

- communes participate in regional decisions on evacuation plans

- communes have veto power over the siting of uranium ore mines and other energy facilities

For Sweden and other countries, it is not clear what the full scope of such risk activities is. There are several reasons to suspect that generic problems may exist-- only limited expertise is available to local communes, there is a tendency to abdicate many risk decisions to regional or national governmental levels, risk issues enjoy only a low visibility in the city or commune, communities must rely on outside consultants or higher governmental expertise.

There is need for an initial study to define the scope of the present community risk management function in industrial societies and to define relevant problems. Specifically, the objectives of the proposed project are:

- to determine what risks are now handled by local communities in several advanced industrial societies

- to identify generic problems or weaknesses in the performance of local communities as risk managers

- to define existing problems in intergovernmental cooperation in risk management

- to suggest a research strategy for a full analysis of the performance of local communities

The suggested project is thus essentially a problem-finding effort. Several approaches may be suggested as starting points--a brief field study of several local communities in selected countries to detail the range of risks treated, an analysis of community development plans to assess the extent to which risk considerations are included, and interviews with local officials responsible for risk control. Particular attention should also be given to assessing the technical and medical expertise available to, and used by, local officials in risk questions.

International Issues in
Risk Management

Project #7: Assessing the "Big" Energy Assessments

The past decade has seen a remarkable number of "big"
energy risk assessments. Each represents a substantial scientific undertaking and involves major commitments of technical and financial resources. The most notable of these
studies include:

The Reactor Safety Study (WASH 1400).[23] In 1975, the
U.S. Nuclear Regulatory Commission published its long
awaited assessment of reactor risks, an effort which involved some $4 million, a study team of 60 persons, 70
person years of effort, a review of the draft by some 90
groups and individuals who provided 1800 pages of comment,
and three years to complete. Using fault tree and event
tree analysis, the study assessed the risks of core meltdowns as 1 per 20,000 reactor years, but the risks of an accident's producing 100 or more fatalities were estimated
as 1 in 10 million per reactor year. While the study has
been widely criticized and the Lewis Report[24] found it
"inscrutable" and recommended that its absolute quantitative values of risk not be used uncritically as the basis
of regulatory decision-making, it is clearly a benchmark
study in nuclear reactor safety.

The Inhaber Report.[25] This controversial Canadian study
sought to provide a comparative assessment of the full spectrum of risks presented by alternative energy fuel cycles,
including the manufacture of materials and the disposal of
wastes. The report also treated both occupational and public risk. The findings argue that the risks of nuclear
power compare very favorably with those of fossil fuels
(except natural gas), and the risks of solar power are
higher than often assumed. The report has been heavily
criticized, however, particularly by John Holdren.[26]

The Canvey Island Study.[27] A 1978 report by the British Health and Safety Executive assessed the risk of a
petrochemical complex located at Canvey Island on the
Thames River. The site includes a natural gas terminal,
several oil refineries, a storage depot for chemicals and
petrochemicals, an ammonium nitrate fertilizer plant, and
a bottling factory for liquefied petroleum gases. Recommendations presented changes to decrease the risks involved
by cost-effective measures. The methodology, with its
in situ orientation, provides a valuable approach to

localized and concentrated risks arising from complexes of
activity.

KBS 1 and 2. A Swedish governmental committee (AKA)
completed in 1975 a report designed to prove that it was
possible to handle and store the high-level waste in an absolutely safe manner. The Swedish industry performed the
KBS project at an expense of approximately $15 million.
The resulting reports, KBS-1[28] and KBS-2[29] were scrutinized
not only in the normal way but were also submitted to widespread national and international review.[30] An interesting
mediation procedure had two scientists, one optimistic and
one skeptical, work together to assess what KBS had proved
and had not proved.[31]

German Reactor Study.[32] Commissioned in 1976, this
study utilized Rasmussen-type methodology but took account
of the higher limits of redundancy in German nuclear plants
and population densities ten times higher than those in the
U.S. Completed in 1979 by the German Reactor Safety Commission (RSK), the study proved the probability of core
meltdown at one in 10,000 reactor years and, like WASH 1400,
found that small leaks in the main reactor coolant pipes
were likely to be the most frequent initiating events. But
the report also estimated that additional safety precautions should ensure that injury would occur to the public
in only one of 100 such meltdowns. Since one-half of all
accidents involving fatalities would cause deaths beyond
Germany's borders, the study called attention to the need
for international coordination of reactor safety and consequence mitigation policies.

Given the major commitment in a number of nations to
these "big" risk assessments, the time is opportune for
inquiring into the usefulness of such studies. The objectives of the proposed project are:

(1) to determine the extent to which these studies
    actually clarify the risks involved

(2) to evaluate the extent to which they have nar-
    rowed scientific debate

(3) to identify their use by key groups in the
    political process and by the mass media

(4) to identify their impact upon safety and regu-
    latory policy

Each of these objectives poses somewhat different requirements for methods and data sources. Determining the extent of risk clarification will involve a review of preceding technical reports to allow an analysis of (a) new risks identified, (b) revision or greater precision (quantification) in probability or consequence estimation, or (c) identification of unused control measures to ensure safety. These analyses should be supplemented by interviews with both proponents and opponents to identify varying opinions on unresolved risk issues.

This latter exercise will also be useful in evaluating the extent to which scientific debate has narrowed as a result of the risk assessment. Charting the risk issues of contention up to and after the assessment should permit some estimate of the impact of the study. The public statements of adversarial scientists and groups should provide a particularly useful source of information, but should be supplemented by personal interviews with influential persons of various points of view and by analysis of the literature (journals, public statements) of interested parties.

The study should devote considerable attention to evaluating the uses made of such risk assessments by key groups in the political process. Undoubtedly there will be considerable selectivity involved in the adoption of risk assessment findings--what is adopted and what is left out? In many cases risk will not be the preeminent consideration; it will often be overwhelmed by other considerations. Particularly important will be the interaction between risk and associated benefits and the response of various decision-makers. In short, the study should carefully assess the degree to which risk assessments <u>penetrate</u> (if they do) the political system and indicate ways of improving the usefulness of such assessments.

Finally, the identification of impacts upon safety and regulatory policy might well involve the reconstruction of the government's receipt of and action upon the report. This should start at least with the peer-review stage and proceed to the present. Formal considerations and actual policy or regulatory changes should be plotted over time and the extent to which they can be attributed to the risk study evaluated. It will be necessary, of course, to sample among the risk reports--a pairing system of influential and weak studies across several energy areas might well prove useful. The study could be conducted as individual projects within particular countries, or, more effectively, as a large coordinated international effort.

Project #8:   Comparing National Responses to Energy Risks

National debates over energy risks are likely to inten-
sify as comparative assessments of risks become available
as major risk events (such as the Three Mile Island accident, the Norwegian drilling platform catastrophe, the
Amoco Cadiz and Torrey Canyon oil spills) occur. There is
much to be learned from how different countries evaluate
and respond to energy risks. Although wide sharing of individual risk assessments has occurred, little is known systematically about the responses of various countries to energy
risks, the motivations which underlie these responses, their
relative success or failure, and the reasons therefor. As
western industrial societies evolve energy policies, a substantial analysis of the experience of other societies would
provide one valuable type of information for major decisions.

Three principal objectives are recognized for this project:

● to determine the extent to which countries respond
  selectively to energy risks and the reasons for
  such selectivity

● to assess the effectiveness of alternative national
  institutions and managerial strategies in reducing
  or mitigating energy risks

● to suggest specific means or lessons for transferring success and avoiding failure

The proposed study is an ambitious one, involving multiple countries and stages of research. A comparative study
of some five to six countries, selected for their range of
approaches and energy resource situations, is envisioned.
The study would be collaborative and would have two major
components--first, an initial overview of major differences
in national responses to energy risks and the relative
success or failure of different approaches; and, second,
detailed case studies of management of particular energy
risks. The proposed study would also comprise two research
stages: an initial stage to define objectives and study
design, to assemble an international team, and to outline
major historical trends and managerial approaches, and a
second stage of more detailed in-depth work and analysis.

In regard to the national overviews of energy risk response, two initial tasks are suggested. The first would

consist of plotting and analyzing the configuration of effort
allocated by each country over the matrix of energy risks.
Appropriate indicators of effort could include numbers of
research reports, formal risk assessments conducted, governmental budgetary allocations (for research and regulation),
personnel distributions, and enactments of regulations. Historical trends in such effort configurations should be examined and compared.

The second task would focus upon defining the trends in
national objectives and managerial efforts and would address
the motivations that prompted particular studies or control
measures. The impact of major studies and assessments should
be assessed. Major contrasts in institutional structures
would be identified, particularly in regard to authority,
responsibility, and accountability. Finally, the effectiveness of various approaches in controlling risks should
be evaluated, along with the reasons for relative success
or failure.

Institutional, economic, and ideological differences
may be expected to play a significant role. Scientific
information and social judgment may interact differently,
and judgments as to what is scientifically and methodolog-
ically admissible may vary. In the U.S., for example, there
has been a tendency to tighten occupational health standards as more sensitive indicators of harm are found. In
the USSR, by contrast, the tendency has been to proceed
from zero dose and the initial characterization of baseline
physiologic-biochemical parameters in experimental individuals. The standard is then set such that the maximum
permissible concentration is below the lowest level that
causes a statistically significant deviation in extremely
sensitive indicators of behavioral or biochemical responses.

The second stage of research should focus on the research questions defined in the preliminary work. Several
energy risks would be selected for more detailed comparative studies. These might include, for example, air pollution from fossil fuel plants, maritime oil spills, risk of
"soft" energy technologies (e.g., solar) or conservation
efforts, and nuclear accidents. A contextual framework
would need to be developed so that risk response might be
considered in historical and social context. The detailed
case analyses would permit in-depth analyses of the effectiveness of different institutional structures or managerial strategies. The results would permit validation and
extension of generalizations derived from the more general
national overview.

Project #9:   Risk Profiles of Developing Countries

Although considerable natural hazards research has been
conducted on developing societies, scientific attention to
technological risks has centered strongly on advanced industrial societies. Improved understanding is much needed as
to the special problems that confront developing societies
as they deal with the broad spectrum of risks. The notion
of "risk" itself may acquire new meaning within the context
of a developing society. And, of course, value conflicts
may be expected between risk reduction and developmental
needs.

The major objective of a developing country is to accelerate the economic progress of the nation and to facilitate
its entry into the development process. The degree of tolerable risk for any activity relates strongly to the associated development benefits. In confronting technological
risks, the developing country suffers from limited technological and human resources and from a lack of mechanisms
for monitoring, researching, and controlling risks. For
risks associated with planned or deliberate new forms of
economic activity, government bears much of the burden for
identifying and assessing the impacts. Yet presently
there is little mobilization of baseline information to
define trends and priorities in risk management.

Such baseline information involves a first approximation of the sources, composition, and configuration of risk
generation. A first priority is an information base of
actuarial (or statistical) data on various societal risks.
Such risks must be identified and arrayed in useful form.
Ordinarily these societies will have no centralized institution for assembling such information; rather it will be
lodged in various governmental agencies, industries, trade
organizations, etc. The form of these data, in the absence
of any coordination, will make overall or cross-hazard comparison difficult. Even in advanced bureaucratic and industrial societies, there are few systematic comparative data
on the risks facing society.

The objectives of this project are:

● to define major risk dimensions, including both
  natural and technological hazards, by which to
  characterize the risk experience in various
  developing countries

● to develop national risk profiles by which

developing countries may be compared and risk
intervention needs and priorities assessed

A national risk profile for risk management would include information on the source of the risk. Is the risk
one from the natural environment, for example, or does it
emanate from the deployment of technology? What are exposure and/or mortality levels? Is the risk the result of
domestic activity or has it resulted from imported technology or productive processes? To the extent possible
the risks should be categorized in some way useful for
managerial activities (e.g., transportation risks) and
(where possible) trend lines plotted over time. It is
recognized, of course, that monitoring and surveillance
systems cannot be put into place to develop systematic and
comparable data. Short of this, however, a useful task
would be the collection, orderly arrangement, and enlightened
commentary on whatever data can be pieced together.

Ideally, risks should be grouped into broad priority
categories for societal attention and effort. Since reality
will usually settle for something less than the ideal, priorities by sector may well prove to be a more realistic
level of intervention and will probably be in greater accord
with existing risk management institutions. So one could
hope, for example, for risk profiles treating workplace
risks, poverty, agriculture, foodstuffs, etc. This information could be used by developing countries, aid agencies,
and international organizations to help channel risk management resources.

Thus far, the focus has been on the statistical portrayal of the variety of risks facing developing societies.
Beyond this, there is a need to know something of the process by which these risks come into existence. Partly this
entails an understanding of causality as it occurs in natural
hazards or in the adverse side effects of technological
change. Many countries will have limited resources for such
demanding tasks, especially where the risks in question are
chronic and synergistic in nature. But a significant part
of the risks may be technological and domestic in origin.
For this class it is essential to inquire into how risks
are generated and how vulnerability may be reduced. In both
socialist and capitalist societies, development managers
have overriding objectives (growth, production targets) to
meet, yet risks undoubtedly constrain goal realization. The
study results should provide useful baseline data on which
risk management programs can be built.

Project #10:   The Export of Risks

Whereas concern over technological risks appears to mount daily throughout advanced industrial societies, developing countries appear to be much more vulnerable to their effects.   Technologies and new products are adopted in developing societies with a very incomplete understanding of long-term effects.   There is also a considerable time lag between the discovery and control of risk-causing substances in developed countries and the implementation of similar protection.   In fact, there is evidence that manufacturers in developed countries frequently exploit this situation by marketing hazardous products in Third World countries.   There is also a reverse flow problem, as agricultural products contaminated by banned pesticides find their way back to developed societies.   The U.S. Food and Drug Administration, for example, has found in its spot checks of imports of fresh vegetables from Mexico more than 50 different illegal pesticides involved and a total of five percent of all shipments exceeded pesticide limits.[33]

This is well illustrated by world patterns of changes in smoking.   Growth in tobacco consumption is currently greatest in the world's poorest countries, a fact which led the World Health Organization to conclude that if immediate action is not forthcoming, smoking diseases will appear in developing countries before communicable diseases and malnutrition have been controlled.[34]   Cigarettes sold in the Third World usually contain twice as much cancerous tar as those sold in developed countries and generally carry no health warning labels.   Antismoking programs are almost unknown in developing countries, and proposals for controlling the risks have met strong resistance.

Tobacco is, of course, not unique; other examples abound:

- in 1976 at least five Pakistanis died and 2,900 became ill from the careless use of the common pesticide Malathion in areas where barefoot children played

- several million children's pajamas treated with the carcinogenic fire retardant TRIS were shipped overseas after the U.S. Consumer Product Safety Commission forced them off the market

- an anti-diarrhea medicine sold only by prescription in the U.S. because it is fatal in overdoses was sold over the counter in Sudan

- after an intrauterine device was linked to the deaths of 17 women in the U.S. and taken off the market, it was widely marketed overseas

- some 120,000 baby pacifiers which were linked to infant choking deaths in the U.S. were exported to Australia by a U.S. manufacturer after a 1978 ban was enacted by the U.S. Consumer Product Safety Commission

It is not surprising that the United Nations Environmental Programme (UNEP) has condemned the export of hazardous products and called for a halt in sales until the risks can be assessed in developing countries. A recent conference on the International Code of Conduct on the Transfer of Technology, held under the auspices of the United Nations Conference on Technology and Development (UNCTAD), accepted a Tanzanian proposal requiring that governments exporting technology help the receiving country understand and avoid potential hazards.[35] Similarly, the United Nations Commission on Transnational Corporations is working on a code of conduct that would require the exporting company to relay to the recipient country all information available on hazardous products or technology.[36] In the United States, a governmental interagency working group has recommended that export licenses be required for hazardous products and that the importing country be fully informed and raise no objections.[37] Yet, despite the growing concern, there is a dearth of research to guide policy.

The exported risk takes a number of different forms:

(a) hazardous substances directly exported to developing countries

(b) manufacturing processes, adopted in developing countries, that are hazardous to workers and/or the public

(c) equipment that is itself hazardous, especially equipment exported to developing countries because of obsolescence or safety regulations in industrial nations

(d) the transplant of entire manufacturing operations owned by U.S. or multinational corporations, with hazardous consequences for workers and/or the public in developing countries

(e) equipment and material that may not be hazardous

when used properly or in its original setting but
can become so with improper use or in altered conditions (as with agricultural chemicals)

The objectives of the proposed project are:

- to estimate the magnitude of the problem(s) posed
  by the export of risk and to indicate current trends
  and configurations

- to identify for developing countries the characteristics that determine vulnerability to such risks

- to assess the extent and adequacy of responses to
  control the problem in both the exporting and importing countries and in various concerned international organizations

- to recommend additional responses needed for more
  effective control of risks

This project would be best undertaken by a collaborative effort between scholars from developed and developing
countries. An initial extended workshop would be very useful for defining the scope of impact, dimensions of the
problem, and alternative research designs. Some detailed
case studies focussed on the different types of risk export
and treating both ends of the transfer chain (the technology
exporter and importer) would be needed. In particular, the
capability of the exporter to assess the risk as the locus
of experience shifts and the ability of the importer to
recognize the risk would need to be carefully assessed.

Conclusion

The proposed risk management research program is beyond
the resources or mandate of any one funding agency. The
objective of these recommendations is not only to demarcate
significant research needs but to sketch the basic outlines
of priority projects. The program argues for a redirection
of current risk research to a firmer grasp of current processes that operate in society and to a broader international framework of investigation.

References and Notes

1. A.-C. Blomkvist and L. Sjöberg, *Risker och olyckstrap-portering*. [Risks and Accident Reports. Reflections with Reference to Current Cases Reported in Swedish Mass Media] (Risk Generation and Risk Assessment in a Social Perspective Project, Göteborg, Sweden, 1976). In Swedish with an English summary.

2. F. Fagnani, "Le débat nucléaire en France (Acteurs so-ciaux et communication de masse)," (Cordes, Paris, 1977). F. Fagnani and J.P. Pages, "Nuclear Energy and Public Opinion," Unesco Round Table Conference 20-23 November, 1978, Vienna (Unesco, New York, 1979).

3. B. Combs and P. Slovic, "Causes of Death: Biased News-paper Coverage and Biased Judgements," *Decision Re-search Report* 78-8 (December, 1978).

4. E.M. Rogers, "Mass Media and Disasters," *Natural Hazards Observer*, 4, No. 3 (March, 1980), 1-2.

5. T.J. Scanlon, R. Luukko, and G. Morton, "Media Coverage of Crises: Better than Reported, Worse than Necessary," *Journalism Quarterly*, 55, No. 1 (Spring, 1978), 68-72.

6. National Research Council, Committee on Disasters and the Mass Media, *Disasters and the Mass Media* (National Academy of Sciences, Washington, 1980).

7. D. Nelkin, *Technological Decisions and Democracy: European Experiments in Public Participation* (Sage Publications, Beverly Hills, California, 1977).

8. Great Britain, Dept. of the Environment, *The Windscale Inquiry: A Report by the Hon. Mr. Justice Parker, presented to the Secretary of State for the Environment on 26th January 1978* (HMSO, London, 1978); Gorleben International Review Board, *Rede-Gegenrede* (Ministry of Social Affairs, Hanover, Federal Republic of Germany, 1979).

9. D. Nelkin, *Technological Decisions.* . . .

10. *Ibid.*

11. *Ibid.*

12. S. Rippon, "Public Relations Symposium," *Nuclear News*, 23, No. 15 (December, 1980), 80.

13. D. Nelkin and M. Pollack, "The Anti-Nuclear Movement in France," *Technology Review,* 83, No. 2 (December, 1980), 36-37.

14. D.J. Gamble, "The Berger Inquiry: An Impact Assessment Process," *Science,* 143 (1978), 946-951.

15. A.M. Weinberg, "Is Nuclear Energy Acceptable?,"*Bulletin of the Atomic Scientists,* 33, No. 4 (April, 1977), 54.

16. P. Slovic, S. Lichtenstein, and B. Fischhoff, "Images of Disaster: Perception and Acceptance of Risk from Nuclear Power," in A. Perlmutter, O.K. Kadiroglu and L. Scott, eds., *Proceedings of the Second International Scientific Forum on an Acceptable World Energy Future* (Ballinger, Cambridge, MA., 1979).

17. H.J. Otway and M. Fishbein, "The Determinants of Attitude Formation: An Application to Nuclear Power," *Research Memorandum,* International Institute for Applied Systems Analysis (IIASA, Vienna, 1976).

18. P.J. Stallen and R.W. Meertens, "Value Orientations, Evaluations and Beliefs Concerning Nuclear Energy" (Internal Report 77 SO-2; Dept. of Social Psychology, University of Nijmegen, Nijmegen, The Netherlands, 1977).

19. C.H. Green and R.A. Brown, "The Acceptability of Risk: Summary Report," (University of Dundee, Dundee, Scotland, 1978); C.H. Green and R.A. Brown, "Counting Lives," *Journal of Occupational Accidents,* 2 (1978), 55-70.

20. L. Sjöberg, "Strength of Belief and Risk," *Policy Sciences,* 11 (1979), 39-57; O. Svenson, "Risks of Road Transportation in a Psychological Perspective," *Accident Analysis and Prevention,* 10, (1978), 267-280.

21. B. Johnson, *A Propositional Inventory of Case Studies in Hazard Management,* (Background Paper No. 1, Clark University, Center for Technology, Environment, and Development, Worcester, MA., 1979).

22. See M. Winell, "An International Comparison of Hygenic Standards for Chemicals in the Work Environment," *Ambio,* 4, No. 1 (1975), 34-36.

23. U.S. Nuclear Regulatory Commission, *Reactor Safety Study* (WASH-1400; NUREG 75/014; Nuclear Regulatory Commission, Washington, 1975).

24. H.W. Lewis, *Risk Assessment Review Group Report to the U.S. Nuclear Regulatory Commission* (NUREG/CR-0400; Nuclear Regulatory Commission, Washington, 1978).

25. H. Inhaber, *Risk of Energy Production* (Report AECB-1119, Revision 3; Atomic Energy Control Board, Ottawa, 1979).

26. J. Holdren *et al.*, "Risk of Renewable Energy Sources: A Critique of the Inhaber Report" (Energy and Resources Group, University of California, Berkeley, 1979).

27. Great Britain, Health and Safety Executive, *Canvey: Summary of an Investigation of Potential Hazards from Operations in the Canvey Island, Thurrock Area* (HMSO, London, 1978).

28. Kärn-Bränsle-Säkerhet (KBS), *Handling of Spent Nuclear Fuel and Final Storage of Vitrified High Level Reprocessing Waste* (Stockholm, 1977).

29. Kärn-Bränsle-Säkerhet (KBS), *Handling and Final Storage of Unreprocessed Spent Nuclear Fuel* (Stockholm, 1979).

30. National Research Council, *A Review of the Swedish KBS-II Plan for Disposal of Spent Nuclear Fuel*, PTO Subcommittee for Review of the KBS-II Plan, Commission on Natural Resources (National Academy of Sciences, Washington, 1980).

31. N. Abrams, "Nuclear Politics in Sweden," *Environment*, 21 (May, 1979), 6-11, 39-40.

32. Gesellschaft für Reaktorsicherheit, *Deutsche Risikostudie Kernkraftwerke* (Bundesministerium für Forschung und Technologie, Bonn, 1979).

33. "Produce Imported from Mexico Often Tainted by Toxic Pesticides," *Hartford Courant*, 8 April 1980.

34. A. Chacko, "Smoking and the Third World," *World Press Review*, 27 No. 4 (April, 1980), 56.

35. "Policies on Exporting Harmful Products are in the Works," *World Environment Report*, 6, No. 15 (30 June 1980), 1-2.

36. *Ibid.*

37. "White House Delays Curbing Hazardous Product Export," *Hartford Courant*, 12 September 1980, p. 47.

Index